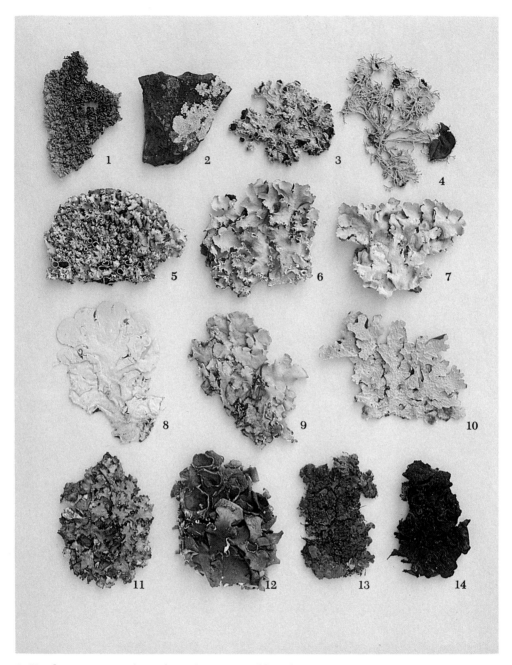

1. *Xanthoria parietina* (p. 31); 2. *Acarospora chlorophana* (p. 31); 3. *Cetraria canadensis* (p. 45); 4. *Letharia columbiana* (p. 98); 5. *Xanthoparmelia cumberlandia* (p. 42); 6. *Parmelia flaventior* (p. 34); 7. *Pseudoparmelia caperata* (p. 34); 8. *Parmotrema linctorum* (p. 66); 9. *Cetrelia olivetorum* (p. 57); 10. *Parmelia sulcata* (p. 78); 11. *Cetraria ciliaris* (p. 134); 12. *Nephroma resupinatum* (p. 131); 13. *Leptogium cyanescens* (p. 154); 14. *Collema nigrescens* (p. 157). Magnification about X 3/4.

how
to
know
the
lichens

The **Pictured Key Nature Series** has been published since 1944 by the Wm. C. Brown Company. The series was initiated in 1937 by the late Dr. H. E. Jaques, Professor Emeritus of Biology at Iowa Wesleyan University. Dr. Jaques' dedication to the interest of nature lovers in every walk of life has resulted in the prominent place this series fills for all who wonder **"How to Know."**

John F. Bamrick and Edward T. Cawley
Consulting Editors

The Pictured Key Nature Series

How to Know the
  AQUATIC INSECTS, Lehmkuhl
  AQUATIC PLANTS, Prescott
  BEETLES, Arnett-Downie-Jaques, Second Edition
  BUTTERFLIES, Ehrlich
  FALL FLOWERS, Cuthbert
  FERNS AND FERN ALLIES, Mickel
  FRESHWATER ALGAE, Prescott, Third Edition
  FRESHWATER FISHES, Eddy-Underhill, Third Edition
  GILLED MUSHROOMS, Smith-Smith-Weber
  GRASSES, Pohl, Third Edition
  IMMATURE INSECTS, Chu
  INSECTS, Bland-Jaques, Third Edition
  LICHENS, Hale, Second Edition
  LIVING THINGS, Winchester-Jaques, Second Edition
  MAMMALS, Booth, Third Edition
  MITES AND TICKS, McDaniel
  MOSSES AND LIVERWORTS, Conard-Redfearn, Third Edition
  NON-GILLED MUSHROOMS, Smith-Smith-Weber, Second Edition
  PLANT FAMILIES, Jaques
  POLLEN AND SPORES, Kapp
  PROTOZOA, Jahn, Bovee, Jahn, Third Edition
  SEAWEEDS, Abbott-Dawson, Second Edition
  SEED PLANTS, Cronquist

SPIDERS, Kaston, Third Edition
SPRING FLOWERS, Cuthbert, Second Edition
TREES, Miller-Jaques, Third Edition
TRUE BUGS, Slater-Baranowski
TRUE SLIME MOLDS, Farr
WEEDS, Wilkinson-Jaques, Third Edition
WESTERN TREES, Baerg, Second Edition

how
to
know
the
lichens

Second Edition

Mason E. Hale

Smithsonian Institution

The Pictured Key Nature Series
Wm. C. Brown Company Publishers
Dubuque, Iowa

Copyright © 1969, 1979 by Wm. C. Brown Company Publishers

Library of Congress Catalog Card Number: 78-55751

ISBN 0—697—04763—6 (Paper)
ISBN 0—697—04762—8 (Cloth)

Printed in the United States of America
10   9   8   7   6

Contents

Preface   vii

What are Lichens   1
   Colors of Lichens   2
   Growth Forms   2
   Isidia and Soredia   8
   Reproductive Structures   8
How to Make Chemical Tests   10
   Color Tests   10
   Microchemical Tests   12
How to Collect and Study Lichens   17
   Collecting Lichens   17
   Uses of Lichens   18
   Projects with Lichens   18
How to Use the Pictured Key   21
General References   22
Pictured Key to the Foliose and Fruticose Lichens   25
   I. Stratified Foliose Lichens   27
   II. Gelatinous Lichens   147
   III. Umbilicate Lichens   158
   IV. Fruticose Lichens   167
   V. Squamulose Lichens   230
List of Synonyms and Incorrect Names   237
Phylogenetic List of Genera and Families   239
Acknowledgments for Illustration   240
Index and Pictured Glossary   241

Preface

Lichens are known to most naturalists as unusual but somewhat mysterious plants. Part of this mystery lies in the fact that there are but few reliable books about lichens written for beginners. Many people collect lichens and are then frustrated when they try to put names on them. Books for identification have been so badly out of date, overly technical, or inaccurate that one has had little chance of correctly naming an unknown specimen.

The first edition of *How to Know the Lichens* was prepared with the aim of filling this gap and it stimulated much interest in lichenology. Now, ten years later, a new edition is needed to incorporate the results of recent taxonomic research and scientific studies. Lichenology still lags far behind the higher plants and even mosses and fungi in terms of easily available reference works. In another sense this means that it is an exciting field because so many new discoveries can be made.

Several more genera have been critically studied since 1969 and can now be identified with greater accuracy, including *Alectoria* and its segregate genus *Bryoria, Collema, Hypogymnia,* and the brown Parmlias and the genus *Neofuscelia*). As a result of these changes and an updating of other groups, 70 new entries have been added to the keys, bringing the total number of species treated to 427, many of them western lichens. The illustrations have been enlarged in many cases in the belief that a close-up at higher magnification is more useful for identification purposes than one at natural size. Another change is a tabulation of the genera and their characters before each major key section. In addition the genera are classified by family phylogenetically at the end of the book.

Crustose forms are still not included because they are far too numerous to present here and too little known. Though imperfect and now out of print, Fink's *Lichen Flora of the United States* continues to be the only overall reference for crusts, especially at the genus level. It can be supplemented by local lists such as Brodo's for Long Island, Weber's for the Chiricahua Mountains of Arizona, and Wetmore's for the Black Hills of South Dakota.

There are, of course, genera that cannot be identified completely at this time. *Usnea,* for example, one of the most easily recognized lichen genera, has not been critically studied and many specimens cannot be named to species. *Ramalina* is also poorly known, and other important genera such as *Heterodermia, Physcia,* and *Physconia* need to be restudied in the light of new discoveries. And finally, as in the first edition, some species known to be very rare or based on dubious records have not been included here because of space limitations.

Many constructive comments on the keys and discussions in the first edition have been included in this revision. Special thanks go to Dr. Robert Egan, Dr. T. L. Esslinger, Dr.

Richard Harris, Dr. P. M. Jørgensen, Dr. Thomas Nash, Dr. Philip Rundel, Mr. Bruce Ryan, Dr. Shirley Tucker, and Dr. C. M. Wetmore, as well as to the 1977 lichen class at the University of Montana Biological Station.

What Are Lichens

Everyone is familiar with plants as green chlorophyll-containing organisms that manufacture their own food. A lichen (pronounced "liken") is also a plant but a very special kind, for when we dissect and examine it under a microscope, we find that it is composed of two completely different organisms, microscopic green or blue-green **algae** that are related to free-living algae and colorless **fungal threads** called hyphae (Fig. 1). These two components grow together in a harmonious association referred to as symbiosis, or more simply a "living together." Lichen symbiosis, however, differs basically from all other kinds in that a new plant body, the **thallus**, is formed, and this thallus has no resemblance to either a fungus or an alga growing alone. This new composite organism behaves as a single independent plant, the green alga manufacturing sugars by photosynthesis and the fungus living off these foodstuffs and making up the bulk of the plant body. The alga and fungus can be separated artificially and cultured in test tubes (see Fig. 24) but most attempts to recombine them in the laboratory to form a new lichen have been unsuccessful.

Lichens are classified as **cryptogams,** a class of lower plants that includes algae (seaweeds), fungi, and mosses. Technically they should be considered as fungi living symbiotically with algae, but in the minds of most lay-

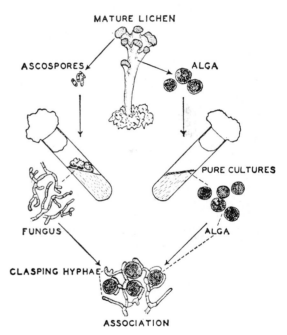

Figure 1 Steps in separating, culturing, and recombining lichen components

men they are combined with mosses which do, in fact, live in similar habitats. True mosses, however, differ in having a green leafy gametophyte that bears a sporophytic capsule (Fig. 2). Lichens lack differentiation into leaves and stems and have distinctive pale greenish yellow or mineral gray colors. The nonlichenized

1

fungi, the cup fungi, edible mushrooms (Fig. 2), bracket fungi, molds, etc., lack a green layer of algae internally and grow on dead trees, leaves, and rotten logs or on humus in dank woods. They are often soft and fleshy and, except for the perennial bracket fungi, usually shrivel when dried or decompose. Lichens, by comparison, most often grow in open or exposed sunny places and appear unchanged when collected and dried.

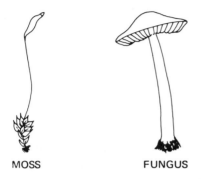

MOSS   FUNGUS

Figure 2   General appearance of a moss and a mushroom fungus

Lichen thalli are generally round in outline, between 1 cm and 30 cm in diameter, and scattered, although clusters of thalli may fuse and cover large areas of tree trunks or rocks. The most easily recognized features of the thallus are growth form and color which together can be used to tell many genera apart in the field. Identification to species can often be done with the naked eye, but some important characters may be too small to be seen without a magnifying lens.

While a stereoscopic binocular scope of 10-30X is ideal for a serious student, such expensive equipment will not be found outside of college or high school laboratories. An ordinary hand lens of 2-10X, costing less than $10, and a good desk lamp will suffice. The desired level of magnification is usually indicated in the keys. Never use more than 10X, even though most binocs go to 20X-30X, unless so instructed. A compound microscope with a magnification of 100-200X is required for anatomical and spore sections and for microcrystal tests.

COLORS OF LICHENS

The colors of lichens are distinctive yet rather subtle and hard to describe. It is extremely important to judge colors only after the thallus is air-dried. If wet or moist, the thallus may turn green. The reason is that the normally opaque upper cortex becomes translucent on wetting, bringing out the green of the chlorophyll-containing algae within the thallus.

The following colors are characteristic for lichens and are shown in the frontispiece: Orange (No. 1, *Xanthoria parietina*), lemon or sulphur yellow to chartreuse (No. 2, *Acarospora chlorophana;* No. 3, *Cetraria canadensis;* and No. 4, *Letharia columbiana*), yellowish green (No. 5, *Xanthoparmelia cumberlandia;* No. 6, *Parmelia flaventior;* and No. 7, *Pseudoparmelia caperata*), whitish mineral gray (No. 8, *Parmotrema tinctorum*), greenish mineral gray (No. 9, *Cetrelia olivetorum* and No. 10, *Parmelia sulcata*), brown (No. 11, *Cetraria ciliaris* and No. 12, *Nephroma resupinatum*), slate blue (No. 13, *Leptogium cyanescens*), and black (No. 14, *Collema nigrescens*).

Color is a very important key character and must be determined with care. The specimen being identified should be matched with the color plate in bright light. With experience it will be possible to judge color at a glance, although one can go so far, for example, as to identify the pigment involved, such as usnic acid in yellowish green lichens, with a microcrystal test or make color tests of the cortex with KOH.

GROWTH FORMS

Growth form means the overall shape and configuration of the lichen thallus. There are three major types: **foliose** (leaf-like), **fruticose** (shrubby or hair-like), and **crustose** (crust-like). A fourth type, the **squamulose** lichens, may also be recognized.

Foliose Lichens

Foliose lichens are flattened and prostrate with an upper surface that is different from the lower, either in color or surface features (Fig. 3). The thallus expands outward from the center as it grows, becoming more or less round in outline. There are definite maximum diameters for these lichens and they can be grouped roughly in size classes as small (1-2 cm in diameter), medium-sized (3-12 cm), and large (13-30). The thallus is usually attached by rhizines over most or all of the lower surface, although suberect species may lack rhizines.

How closely a thallus is attached to a rock or tree bark is an important character that calls for a certain amount of personal judgment that will develop with experience. Appressed or closely adnate thalli (Fig. 3A) cannot usually be peeled from the substrate without damage.

Adnate, loosely adnate, or loosely attached thalli (Fig. 3B) can be removed with a knife. Finally, suberect or ascending species (Fig. 3C) can be removed by hand.

Umbilicate lichens (see Fig. 29) are essentially foliose in growth form but attached to the substrate by a single cord. They may or may not have rhizines in addition to the cord but the outer edge of the thallus is always free.

The typical foliose thallus is divided into numerous branches called **lobes.** These lobes tend to elongate and fork and maintain a more or less constant width. Lobe width, as with thallus color and adnation, is one of the most important characters in the keys. It should be measured carefully with a small plastic ruler marked in millimeters. Do not try to guess at the width. The area measured is ideally just below the lobe tip where a uniform width is reached. Lobes may widen irregularly and overlap, making measurement more difficult. Take an average of several lobes. Narrow lobes (Fig. 4A) are 0.1-3.0 mm wide and tend to be linear or strap-shaped with blunt or angled tips. Broad lobes (Fig. 4B) range from 3 to

Figure 3 Typical examples of foliose lichens: A. adnate (*Physcia phaea*); B. loosely attached (*Parmotrema stuppeum*); and C. suberect (*Pseudevernia intensa*)

20 mm wide and are more irregular in width with rounded or rotund tips.

Margins of lobes are smooth to variously indented, varying from crenate to dentate, dissected, finely divided, or lacerate. **Cilia** are

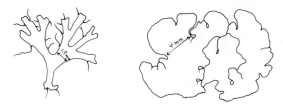

Figure 4   How to measure width of lobes

hairlike structures along the margin that occur in many species (Fig. 5A). They are 0.5-6 mm long and can usually be recognized with a hand lens or by holding the specimen up to a bright light. **Lobules** of various types (Fig. 5B) are also produced in some species. In *Cetraria* tiny black pycnidia are produced on the lobe margins (Fig. 5C).

The upper surface of lobes may be smooth or wrinkled to ridged. The cortex surface, as seen with a hand lens, may be continuous or finely reticulately cracked (Fig. 6A), with white pores (Feb. 6B), with white markings (Fig. 6C), with white pruina (Fig. 6D) so characteristic of *Physconia,* with white spotting (Fig. 6E), or with wrinkling (Fig. 6F) as seen in *Leptogium.* Small warts on the surface of *Peltigera aphthosa* are called cephalo-

dia (Fig. 6G); they contain blue-green algae. Saclike cephalodia are known in some species of *Stereocaulon.*

The lower surface of the lobes has a number of structures that are useful in classifying lichen species. The color varies from jet black to some shade of brown, buff, tan, or ivory to white. Specimens with a black lower surface may have a narrow brown or even mottled brown-white zone at the tips. If the thallus lacks a lower cortex, the white cottony fibrous medulla is easily seen (Fig. 7A). If a cortex is present the surface is smooth, shiny, and sparsely to densely covered with rhizines or tomentum.

Rhizines are compacted strands of fungal hyphae, produced from the lower cortex, which attach the thallus to the substrate. They may be simple (Fig. 7B), once or twice furcate, or more elaborately branched, either dichotomously or squarrosely (see glossary). Branching patterns should be determined with a hand lens.

Tomentum (Fig. 7C) consists of individual hyphae rather than strands as in rhizines (see Fig. 8). It forms a pale brown to black felty mat over the lower surface. Beginners will often confuse tomentum and rhizines but with practice these structures can be recognized with a hand lens.

Distinct pores occur in the lower surface of several genera that have tomentum. Unique

Figure 5   Structures on margins of lobes: A. cilia (*Parmotrema crinitum,* ×1); B. lobules (*Heterodermia squamulosa,* ×10); and C. pycnidia (*Cetraria ciliaris,* ×15)

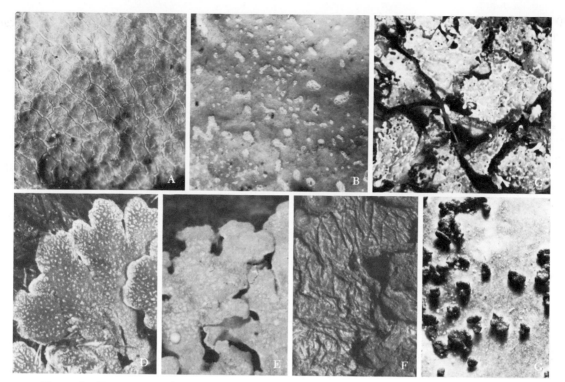

Figure 6   Structures on the upper surface of lobes: A. reticulate cracking (*Parmotrema reticulatum,* ×10); B. pores (pseudocyphellae of *Cetrelia chicitae,* ×10); C. white markings (*Parmelia omphalodes,* ×3); D. white pruina (*Physconia muscigena,* ×10); E. white-spotting (*Physcia aipolia,* ×10); F. wrinkling (*Leptogium phyllocarpum,* ×); and G. cephalodia (*Peltigera aphthosa,* ×10)

cyphellae (see Fig. 85), sunken pits scattered among the tomentum, can be seen without the aid of a lens; they are characteristic of *Sticta.* Another type of pore (**pseudocyphellae**) is

Figure 7   Structures on the lower surface of lobes: A. ecorticate and fibrous (*Heterodermia casarettiana,* ×10); B. rhizines (*Parmelia saxatilis*); and C. tomentum (*Coccocarpia cronia*) (about ×15)

smaller and represents simply a hole in the cortex filled with white hyphae. It is always found in the genus *Pseudocyphellaria* (Fig. 85) as well as in the upper cortex of *Cetrelia* (see Fig. 6B), and *Parmelia.*

The internal structure of foliose lichens must be examined under a microscope. Free-hand sections can be made with a razor blade after some practice or a freezing microtome can be used. Ordinary histological staining procedures may be followed to prepare permanent slides. Safranin, fast green, orcein, and other stains work well. The interior of the thallus in **stratified** lichens consists of an upper **cortex** of compressed cells, a thin but distinct **algal layer** below this, a thick loosely packed **medulla** composed of hyphal strands, and a lower cortex which may be lacking in a few genera (Fig. 8). The usual alga is a green unicell, *Trebouxia*, more rarely a blue-green such as *Nostoc*.

Unstratified foliose lichens have no differentiation into an algal layer and medulla. Instead the algae (*Nostoc, Stigonema,* and other blue-greens) are intermingled with hyphae in a dark, heavily gelatinized layer (Fig. 9). Only *Leptogium* has a primitive cortical layer. These lichens are usually dark brown, black, or bluish slate-colored. They are called gelatinous because the thallus swells and becomes somewhat jellylike when wet.

Fruticose Lichens

Fruticose lichens (Fig. 10) consist of simple or divided branches that are round to flattened in cross section but with little difference be-

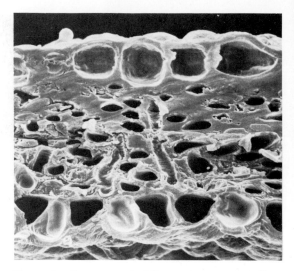

Figure 9 Cross section of *Leptogium azureum* showing upper and lower cortex one cell thick and the unstratified medullary area (scanning-electron microscope view) (×1000)

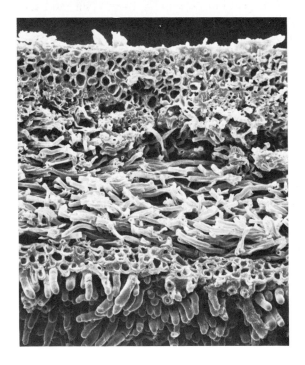

Figure 8 Internal structure of a foliose lichen viewed with a scanning-electron microscope (*Sticta weigelii*), showing upper cortex, medulla, lower cortex, and layer of tomental cells (×400)

tween the upper and lower surfaces. Rhizines are lacking and cilia extremely rare. These lichens are bushy, hairy, or strap-shaped, attached at the base to trees or rocks or in some genera such as *Cetraria, Cladina, Dactylina,* and *Thamnolia,* free growing on soil. Growth occurs at the tips of branches and some species may grow to 2 meters long.

When a fruticose lichen is sectioned and examined under a microscope, one will find a cortex, a thin layer of algae below this, and a medulla. The center may be solid or hollow. The genus *Usnea* is unique in having a dense central cord. *Cladina* is unusual in lacking a distinct cortex. The variety of internal organization is much greater in fruticose lichens than in the foliose forms.

Squamulose Lichens

The thallus of squamulose lichens consists of small, separate, lobelike structures that have an upper cortex, algal layer, and a medulla, but lack a lower cortex and rhizines. They are characteristic of the primary thallus of *Cladonia*

Figure 10 Typical fruticose lichens: A. *Ramalina farinacea;* B. *Cladonia ecmocyna;* and C. *Ramalina menziesii*

(Fig. 11) but also occur in some species of *Dermatocarpon* and in *Psora* (see Fig. 472). While the squamules are only 1-10 mm long they may form clumped mats that cover large areas of soil.

Crustose Lichens

Although true crustose lichens as a group are not treated in this book, several conspicuous marginally lobate species of *Acarospora* (Frontispiece No. 2), *Caloplaca, Candelina, Dimelaena, Placopsis,* and *Rhizoplaca* are included. In these species the central part of the thallus is areolate or chinky. Except for the lack of a lower cortex and rhizines, they come very close to the appressed foliose types such as *Physciopsis syncolla.* True crustose lichens may also have a fairly thick thallus but the margins are unlobed and sometimes fade into the substrate and become indistinct (Fig. 12). It is impossible to remove them from the bark or rock without destroying the thallus. One crustose lichen deserves special mention. This is *Lepraria finkii,* a fragile white powder that is found very frequently at tree bases or in shaded nooks on rocks.

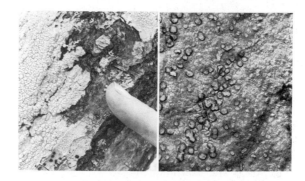

Figure 12 Typical crustose lichens: *Pertusaria multi-puncta* (left) and *Huilia (Lecidea) albocaerulescens* (×3)

Figure 11 Squamulose primary thallus of *Cladonia*

ISIDIA AND SOREDIA

The two most diagnostic characters for all lichen groups are **isidia** and **soredia**, which are vegetative propagules unique to lichens. They must be recognized without hestitation and distinguished clearly with a hand lens.

Isidia are fingerlike cylindrical to flattened outgrowths from the upper cortex, scattered more or less evenly over the upper surface (Fig. 13). Since they are only 0.3-1.0 mm high, they must be examined with a hand lens. They may be simple or branched, sparse to dense, papillate, globular, coarse to granular, even somewhat sorediate, or dorsiventral and flattened. Isidia break away from the thallus, leaving a scar, and may regenerate into a new lichen.

Figure 14 Examples of soredia: orbicular soralia of *Parmelia subrudecta* (left) and linear soralia of *Cetraria oakesiana* (×5)

Soralia may be linear along the lobe margins or orbicular on the surface or tips. Diffuse soredia not organized into definite soralia form on the whole of podetia in many Cladonias.

REPRODUCTIVE STRUCTURES

The most easily recognized reproductive structures are **apothecia,** cup- or disc-shaped bodies, 1-20 mm in diameter. These occur on the upper surface (Fig. 15), along lobe margins as in *Cetraria,* on the lower surface of tips as in *Nephroma,* or at the tips of podetia as in *Cladonia* (Fig. 15). The disc is usually some shade of brown, or more rarely orange, red, or yellow. The disc, when sectioned, shows a uniform layer of sterile threadlike **paraphyses** and scattered asci under the microscope. Each **ascus** contains 1-8, rarely more, **spores** (Fig. 16). The most important characters of spores are number of cross walls (septation) and color, which separate lichens into genera. In practice, however, when identifying foliose or fruticose lichen species, we do not use spores very often.

A second kind of reproductive structure is the **perithecium,** a flask-shaped structure

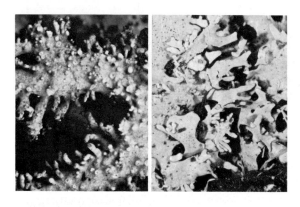

Figure 13 Examples of isidia: *Pseudevernia consocians* (right) and *Hypotrachyna prolongata* (×10)

In contrast, soredia originate in the medulla and erupt at the lobe surface as a powder which is easily brushed away (Fig. 14). When this powder is examined under a microscope, we see clumps of a few algae closely surrounded by hyphae forming a round soredium about 50 microns in diameter. Masses of soredia, called soralia, are usually visible on the thallus surface with the naked eye but in doubtful cases a hand lens should be used.

Figure 15 Reproductive structure of lichens: A. apothecia of *Physcia aipolia* (×10); B. apothecia and pycnidia (black dots) on *Parmelia bolliana* (×3); and C. podetia and apothecia of *Cladonia clavulifera* (×2).

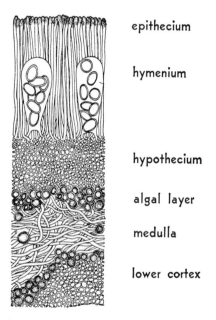

epithecium

hymenium

hypothecium

algal layer

medulla

lower cortex

Figure 16 Cross section of a typical apothecium

buried within the thallus of *Dermatocarpon*, a foliose lichen, and of many crustose genera. It appears at the upper surface as black dots (see Fig. 309). When sectioned, perithecia will also be found to contain paraphyses and asci with spores. Very similar structures called pycnidia are common on the upper surface of foliose lichens (Fig. 15B). They contain numerous, free microconidia, only a few microns long and often difficult to see with a student microscope. Fortunately there is rarely any need to look for pycnidia.

The reproduction of lichens in nature is a mystery. Sexual reproduction in which spores germinate and recombine with algae is theoretically possible but no one has been able to follow these steps in nature. Soredia, isidia, and thallus fragments can all act as vegetative propagules and when dislodged apparently resume growth to form a new thallus. Whatever method is used lichens are eminently successful colonizers in nature.

How to Make Chemical Tests

Chemistry is not an important character in most plant groups, but in lichens it is a useful and practical means of identifying species.

Lichens produce unique chemical substances, most of them weak phenolic or fatty acids which are deposited on the surface of the hyphae. The main classes of substances are shown in Fig. 17. Each lichen usually has a constant chemical makeup so that the same species will give the same test, no matter where it was collected. This means that we can identify lichens more accurately by checking the chemical composition. Moreover, species that are easily confused without careful study, such as *Physcia aipolia* (medulla KOH+ yellow) and *P. stellaris* (medulla KOH−), are quickly separated by simply applying a drop of KOH to the medulla of unknown specimens. Some species, however, that are identical in appearance, have different chemistry. These "chemical species," which can be separated only by a chemical test, are mentioned in the keys where they occur.

COLOR TESTS

A color test is made simply by applying a drop of reagent on the thallus surface or exposed medulla. If the test is positive, there will be a rapid color change, usually red or yellow; if negative, nothing happens. Ideally the tests should be done under a low-power stereoscopic scope, leaving both hands free to apply the reagent, but a hand lens will be satisfactory with practice.

Three different reagents are used: calcium hypochlorite (bleaching powder, abbreviated C), potassium hydroxide (caustic lye, abbreviated K or KOH), and paraphenylenediamine (P). Calcium hypochlorite has now been replaced in everyday use by liquid bleaches (Clorox, for example) which have sodium hypochlorite as the active ingredient and are more stable. In any event, a C solution should be tested periodically on *Parmelia rudecta, Parmotrema tinctorum,* or other common C+ red lichen to be sure it is still active.

Potassium hydroxide (KOH or abbreviated K) comes as dry sticks that should be stored in tightly closed bottles since they absorb moisture. Small pieces can be chipped off and dissolved in water in a small vial. Be sure to make a concentrated solution. This reagent is stable for several months. KOH can be purchased in drug stores or chemical supply houses; it is caustic and must be handled with care. An alternative solution is "liquid plumber" sold in drugstores or hardware stores. It can be used directly from the container.

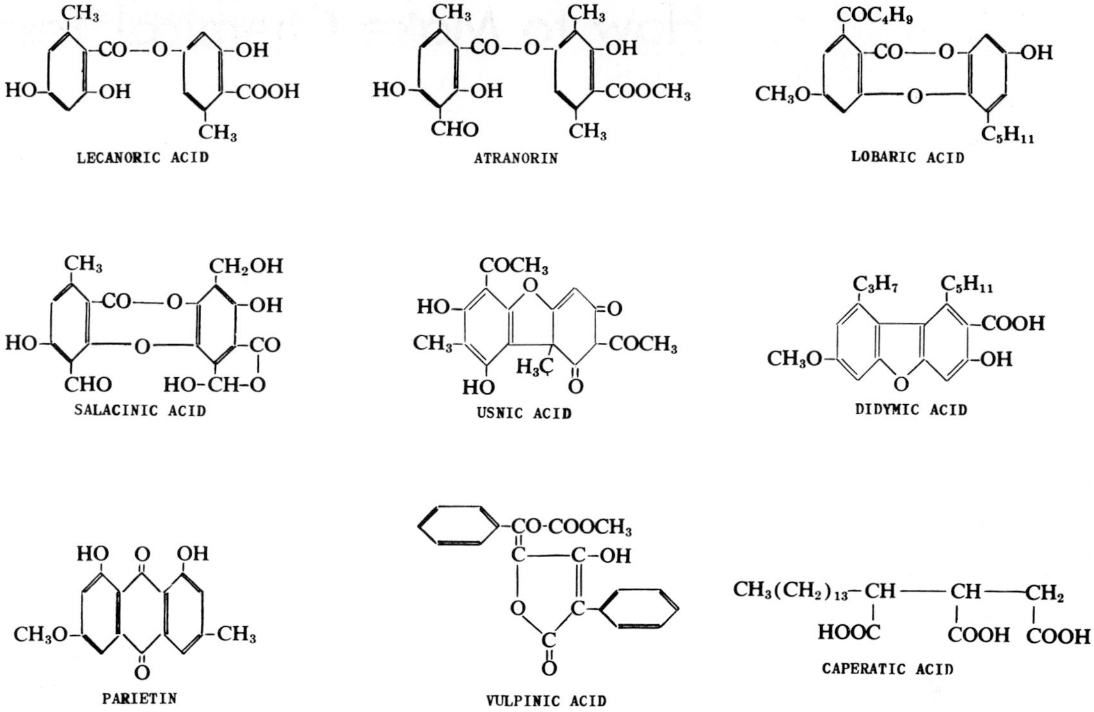

CH₃ structures...

LECANORIC ACID — CH₃ / -CO—O- / -OH / HO- / -OH / -COOH / CH₃

ATRANORIN — CH₃ / -CO—O- / CH₃ / -OH / HO- / -OH / -COOCH₃ / CHO / CH₃

LOBARIC ACID — COC₄H₉ / -CO—O- / -OH / CH₃O- / -O- / C₅H₁₁

SALACINIC ACID — CH₃ / CH₂OH / -CO—O- / -OH / HO- / -O- / CO / CHO / HO-CH-O / O

USNIC ACID — COCH₃ / O / HO- / =O / CH₃- / -COCH₃ / H₃C / HO / O

DIDYMIC ACID — C₃H₇ / C₅H₁₁ / -COOH / CH₃O- / -OH / O

PARIETIN — HO O OH / CH₃O- / -CH₃ / O

VULPINIC ACID — CO-COOCH₃ / C—C-OH / O / C / O

CAPERATIC ACID — CH₃(CH₂)₁₃—CH——CH—CH₂ / HOOC COOH COOH

Figure 17   Molecular structures of some common lichen substances

Paraphenylenediamine is a dark powder prepared for the tests by dissolving a pinch in 5-10 ml of ethyl alcohol (denatured alcohol or rubbing alcohol or even acetone, if nothing else is available). This reagent can only be purchased at a chemical or medical supply company. It must also be handled with great care because the spilled solution or powder will discolor and ruin clothing and paper. The powder should not be inhaled since amines are carcinogenic. Many students will have difficulties purchasing this reagent and the keys are designed to require as little use of it as possible.

Color tests should be done with a hand lens. Carefully scrape away part of the upper cortex with a razor blade to expose an area of medulla about 2-3 mm square (Fig. 18). Apply the reagent with a thin pipette or fine medicine dropper and note any color change as the reagent is being applied. Reagents are applied directly on the thallus surface when a cortical

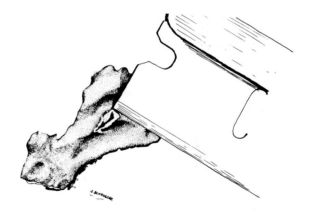

Figure 18   How to scrape the cortex away for a color test on the medulla

test is called for. The keys will indicate which layer has to be tested. When K is applied to the cortex, a spurious yellowish color will sometimes develop after 5-10 seconds because the underlying algae become moist. This false reaction must be carefully distinguished from the immediate K+ yellow test caused by atranorin. When there is doubt, a microchemical test will decide the issue by definitely showing the presence or absence of atranorin or other lichen substance.

MICROCHEMICAL TESTS

It is entirely possible to identify lichens without ever making a microchemical test, and indeed most lichenologists use only the color tests. The keys in this book were constructed specifically to avoid involved tests. The value of microchemical tests, however, lies in their use to make positive identification of acids causing a particular color reaction. For example, since lecanoric and gyrophoric acid both react C+ red, only a microchemical test will distinguish them accurately. A serious student will not long be satisfied simply to make color tests. The notes below provide a brief introduction to microchemical methods. Unfortunately there is no complete summary on this subject. Most of the information has been published piecemeal in journals that are difficult to find. Brief summaries are presented in Hale's *Biology of Lichens,* in Thomson's *Cladonia* book, and in Taylor's *Lichens of Ohio.*

Crystal Tests

In a crystal test, the acid is dissolved from small fragments of the thallus with acetone, and the crude residue that remains is recrystallized from various reagents on a microscope slide. The reagents in common use are abbreviated as follows and mixed in the volume ratios indicated:

> G.E. (glycerin-acetic acid, 3:1)
> G.A.W. (glycerin-95% alcohol-water, 1:1:1)
> G.A.o-T. (glycerin-alcohol-o-toluidine, 2:2:1)

These reagents are conveniently stored in brown dropper-bottles of 50 to 100 ml capacity and except for G.E. and G.A.W. should be prepared fresh every 3 to 6 months. o-Toluidine may be obtained from chemical supply houses. Acetone is sold in drugstores.

Fragments of the lichen thallus are heaped in the center of a microscope slide, and drops of acetone are added several times (Fig. 19). After the acetone evaporates, there should be a whitish or yellowish powdery (rarely gummy) ring of residue. (If no residue forms, do not attempt any tests.) The thallus fragments are carefully brushed away, a small drop of reagent put on a coverslip, and the coverslip placed over the residue. The slide is gently heated over an alcohol lamp, low Bunsen flame, or match until bubbles just begin to form. On cooling a few minutes, crystals will begin to form first around the undissolved residue, later at the perimeter of the coverslip. The shape and color of the crystals are determined under a low-power microscope (100X), and the crystals identified by comparison with photographs. A key to the more common acids

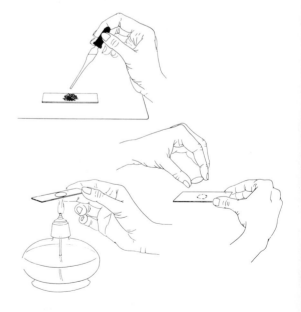

Figure 19 Steps in making a microcrystal test

is outlined below with their color reactions and crystal shapes, where appropriate. Those without a crystal test indicated should be examined with chromatography. It is best to practice first with lichens of known or proven composition.

Pigments

Orange or red pigments, K+ purple: Parietin, skyrin, solorinic acid.

Yellow pigments, K− (or yellowish): Calycin, pinastric acid, pulvic acid, usnic acid (straight yellow needles in G.E., Fig. 20A), vulpinic acid.

Colorless Substances

K+ yellow or yellow turning red: Atranorin (straight needles in G.E., yellow curved needle clusters in G.A.o-T., Fig. 20B), baeomycic acid, galbinic acid (deep yellow warts in G.A.o-T. scattered on atranorin crystals), norstictic acid (4-angled yellow lamellae in G.A.o-T., Fig. 20C), physodalic acid, salacinic acid (yellow orange boats in G.A.o-T., Fig. 20D), stictic acid (colorless hexagons in G.A.o-T., Fig. 20E), and thamnolic acid (yellow fascicles in G.A.o-T. with bubbling).

K− (or brownish), P+ yellow, orange, or red: fumarprotocetraric acid, pannarin, protocetraric acid (yellow warts in G.A.o-T., Fig. 20F), psoromic acid.

K−, P−, C+ pink or red: Anziaic acid, gyrophoric acid (warts in G.E., Fig. 20G), lecanoric acid (curved needle clusters in G.A.W. and G.E., Fig. 20H), olivetoric acid (gummy residue, long curved needles in G.A.W., Fig. 21A), scrobiculin.

K−, P−, C+ green: Didymic acid, strepsilin.

K−, P−, C−, KC+ pink or red: Alectoronic acid (gummy residue, small fan-shaped lamellae in G.A.W., Fig. 21B), cryptochlorophaeic acid (long curved needle clusters in G.A.W., Fig. 21C), glomel-liferic acid, lobaric acid, norlobaridon, physodic acid (short curved needle clusters in G.A.W., Fig. 21D).

K−, P−, C−, KC−: Barbatic acid (prisms in G.E.), bellidiflorin, caperatic acid (feathery globules in G.E., Fig. 21E), diffractaic acid, divaricatic acid (crisscrossed needles in G.A.W., Fig. 21F), evernic acid (bushy clusters in G.E., Fig. 21G), grayanic acid (long straight needles in G.E., Fig. 21H), homosekikaic acid, lichexanthone (orange fluorescent in UV), merochlorophaeic acid (long lamellae in G.A.W. and G.E.), perlatolic acid (gummy residue, long curved needles in G.A.W.), protolichesterinic acid (feathery lamellae in G.E.), rangiformic acid, sphaerophorin, squamatic acid, tenuiorin (long curved needles in G.A.o-T.), ursolic acid, zeorin (hexagonal prisms in G.A.o-T.).

The crystal tests fail most often because (1) too much reagent is added and the coverslip floats, (2) overheating dissolves all of the residue, and (3) the residue is too small to recrystallize. Undissolved residue or sand grains are sometimes mistaken for crystals; only clean crystals free of debris should be examined.

Chromatography

The crystal tests have proved to be reliable and consistent means for identifying most of the lichen substances. However, some substances form no crystals, when crystallized are not distinctive, or fail to crystallize because of interference from other substances present. These difficulties have been largely overcome with paper and thin-layer chromatography. Lichen substances can be chromatographed easily with standard one-dimensional techniques that are now commonplace in most laboratories (Fig. 22). Rather than rely on R_f values, which vary depending on solvents and temperature,

Figure 20 A-H A, usnic acid from G.E.; B, atranorin from G.A.o-T; C, norstictic acid from G.A.o-T.; D, salazinic acid from G.A.o-T.; E, stictic acid from G.A.o-T.; F, protocetraric acid (warts) with atranorin from G.A.o-T.; G, gyrophoric acid from G. E.; and H, lecanoric acid from G.A.W.

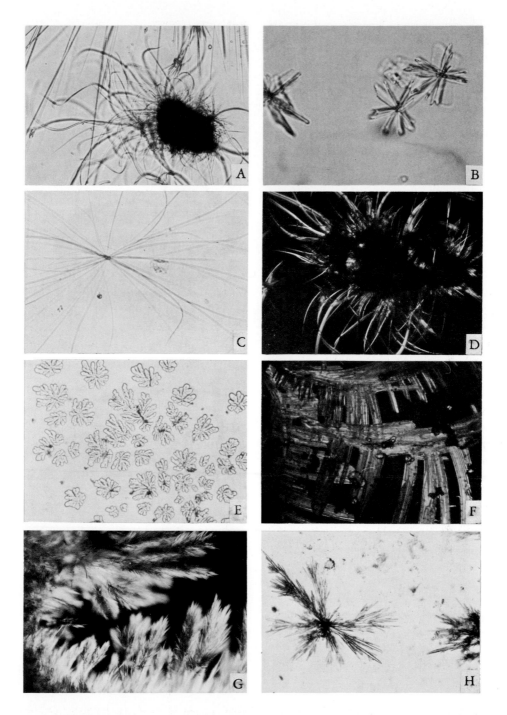

Figure 21 A-H   A, Olivetoric acid from G.A.W.; B, alectoronic acid from G.A.W.; C, crypto-chlorophaeic acid from G.A.W.; D, physodic acid from G.A.W.; E, caperatic acid from G.E.; F, divaricatic acid from G.A.W.; G, evernic acid from G. E.; and H, grayanic acid from G.E.

one should always run known compounds along with the unknowns on the same chromatogram and compare the spots. It will take some practice to interpret spots on chromatograms, but by using the method of comparing spots and using different solvents, unknowns can eventually be identified with detective work.

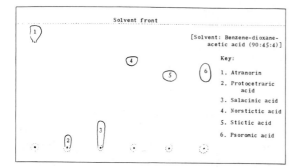

Figure 22 Tracing of a thin-layer chromatographic sheet to show relative positions of lichen substances

An acetone extract of the thallus is applied to the sheet or precoated plate with a micropipette or fine capillary tube, blowing to keep the spot small. The plate is then put into a chamber containing a solvent such as toluene-dioxane-acetic acid (90:45:4), or any of several others. Experimentation will determine which solvent system gives best resolution of spots. Colorless substances will not be visible after the solvent rises on the plate, but they can be visualized by spraying the sheet with 5% sulfuric acid and heating at 100°C for 10-15 minutes. Each acid has a characteristic color. More details on this technique can be found in Culberson's text on lichen products.

Fluorescence Analysis

Students with access to an ultraviolet lamp, even an ordinary sun lamp with appropriate filters, will find that some lichens show brilliant fluorescence. Bright white fluorescence is characteristic of squamatic acid, which is an important diagnostic acid in *Cladonia*. In the foliose groups alectoronic acid is highly fluorescent; divaricatic acid, evernic acid, and perlatolic acid fluoresce noticeably but with less intensity. Lichexanthone and rhizocarpic acid, both in the cortex of some lichens, turn a brilliant orange. Be extremely careful to shield your eyes from the ultraviolet rays; serious eye damage can result, even from radiation reflected from the surface of white chromatograms exposed under a UV lamp.

How to Collect and Study Lichens

COLLECTING LICHENS

Lichens grow on trees, dead wood, rocks, tombstones, mosses, soil, and other substrates. Each species will usually grow best on only one kind of substrate, that is, a lichen which grows on trees normally will almost never be found on rocks, although some very common species such as *Parmelia rudecta* and *Parmelina aurulenta* occur on rocks and trees. The vast majority prefer sunny, exposed habitats, but a few genera, especially those with blue-green algae such as *Collema, Leptogium,* and some Peltigeras, often grow best in moist, shady woods. Some species of *Dermatocarpon* and *Leptogium* grow on rocks near running water or lakes but only *Hydrothyria venosa* actually grows submerged.

Ideal collecting sites for lichens are open oak woods, talus slopes, and rock outcrops throughout the Appalachian Mountains, Ohio River Valley, and southern United States. Some of the most interesting species are found in the top branches and canopy, and a thorough collector will always be on the lookout for recently felled trees. For sheer abundance the spruce-fir and pine forests of northern United States and Canada, the Pacific coast, and the Cascades are unrivaled. Deserts will have many soil and rock species, but a surprising number of species will be found on scrub oaks, junipers, etc., in the Southwest. The least promising areas for lichens are the heavily farmed Midwest and Prairie states, many river bottom forests in the Central States, and forests adjacent to large metropolitan areas. Air pollution is slowly decimating urban lichen floras everywhere.

Lichens are usually collected with a knife or hammer and chisel. The thalli are fairly firm and leathery, but some species become extremely brittle when dry and crumble easily. Since all lichens turn pliable and rubbery when moistened, it is often wise to wet the thalli before attempting to collect specimens which adhere closely to rock or bark. A specimen as large as the palm of the hand should be collected if at all possible. The important thing to remember is that the larger and more uniform a collection is the more scientific value it has. Specimens can be put in paper or cloth sacks if dry and stored indefinitely. Wet specimens should be spread out on newspaper to dry. *Never put lichens in plastic collecting bags* that are now used so often for fungi or higher plants. Moisture cannot evaporate from these bags and the lichen thalli, even if just slightly moist, will quickly discolor or mold.

Dried field collections may be sorted and curated at leisure. Small specimens on bark need only be trimmed somewhat and pasted on 3 × 5 cards. Larger foliose and fruticose specimens often require wetting and light pressing between blotters and corrugates if they are too bulky. Lichens are best dried in front of a fan or by changing to dry blotters several times. Uncirculated artificial heat must be avoided because of the danger of molding and color changes. Specimens are finally placed in standard paper packets (Fig. 23) and a label attached. The packets may be stored upright in shoe boxes or pasted on herbarium sheets and stored in genus folders in herbarium cases.

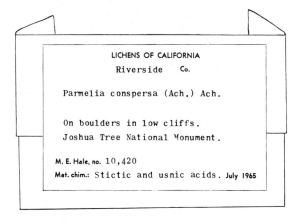

Figure 23 Example of a folded packet (using 8½″ × 11″ paper) and a label

As a final note, you should be sure to obtain permission before collecting lichens in State and National Parks. Do not completely collect or destroy small or rare colonies of lichens, but leave some in place in order to continue the species. Lichens grow very slowly and can be exterminated by overcollecting.

USES OF LICHENS

The main economic importance of lichens is in their use as antibiotics. In Europe yellow lichens (mostly *Cladonia*) are harvested for the extraction of usnic acid, which is the base of an effective antibiotic salve. Lichens, particularly Iceland Moss (*Cetraria islandica*), are still used as expectorants and have an important place in Chinese medicine. Desert tribes in North Africa smoke lichen mixtures.

Other lichens in Europe are extensively collected, and the essential oils they contain used as a perfume base and fixative. Some serve as a mash for making alcoholic beverages. Litmus paper, a familiar acid-base indicator in chemical laboratories, was originally made with amphoteric lichen dyes.

Lichens have some food value, according to reports as much as breakfast cereals, and can be eaten in emergencies. Rock tripes (*Umbilicaria*), for example, are boiled to extract a gelatinous soup thickener. They can also be washed with cold water to remove dirt particles and eaten directly or fried in fat. Lichens containing fumarprotocetraric acid ("Iceland Moss," many Cladonias) should be avoided because they have a bitter taste.

The Reindeer Mosses (*Cladina rangiferina, C. evansii*, etc.) are widely employed as facsimile trees and shrubs in model train or city layouts, miniature Japanese gardens, and in floral arrangements. They are treated with glycerine to make them pliable and dyed green or red.

Rocks with attached lichens may be added to rock gardens. Some may succumb to air pollution or a change in locality, but they can add color and variety for a number of years.

PROJECTS WITH LICHENS

There are many interesting projects that can be done with lichens. If carried out with care, they can add to our scientific knowledge about lichens.

Lichens and Air Pollution

Lichens are extremely sensitive to air pollution, especially sulphur dioxide. Their absence may be used as a measure of how much a city is polluted. For example, if one methodically collects lichens in and around his own city, a map of the species distribution will show the patterns of air pollution. The sooner projects like this are started the better, for levels of pollution are rising. Resurveys several years later can show how fast and in what direction pollution is progressing.

Lichen Dyes

Lichens were used by American Indians and are still the basis of cottage dye industries in northern Europe. Bolton's excellent little text summarizes the techniques used. Briefly, one can extract lichen dyes by boiling a panful of thalli in water for several hours, using any of the common lichens, or by treating C+ red lichens with ammonia for a week or two without heat. White wool dipped or soaked in the extracts will turn various shades of red, orange, or brown, depending on the length of time it is left in.

Culturing Lichens

Physiological studies of lichens can only be conducted if one has access to an autoclave for sterilizing glassware and media. Ahmadjian's text on symbiosis contains information on culture techniques. Spore isolation from moist apothecia is preferred with culture in test tubes (Fig. 24) or flasks. Some crustose species form good colonies in a few weeks, but others may never germinate or grow only very slowly. Presence or absence of antibiotic substances can be tested in these cultures.

Lichen Growth

Lichen students interested in photography will find colorful and unusual subjects among

Figure 24   Test tubes containing isolated lichen fungi (mycobionts) growing on agar

lichens, and they can also put this hobby to scientific use by following the growth rates of lichens. Take a medium-sized foliose lichen and photograph it with a closeup lens along with a ruler. Photograph it again after a few weeks and measure how much growth has occurred (Fig. 25). Variation from month-to-month and summer-to-winter can be studied. Much work of this type remains to be done and real contributions can be made here.

Figure 25 Examples of photographs for recording growth rates (a colony of *Pseudoparmelia baltimorensis* taken at a two year interval)

How to Use the Pictured Key

This key contains 427 main entries and mentions an additional 250 rarer species or chemical variants. These are divided into five main sections: I. Stratified Foliose, II. Gelatinous, III. Umbilicate, IV, Fruticose, and V. Squamulose, with the keys beginning with couplet number 1 in each section. The number in parentheses in the first couplet indicates the couplet from which it originated so that if one becomes lost it is possible to trace back to an earlier couplet.

Before attempting to identify specimens, you should read the introductory chapters in order to familiarize yourself with the new terminology. You should also have ready a hand lens or stereo binocular, a single edge razor blade, and small vials of at least KOH and C (Clorox). Read carefully and fully each pair of couplets and go on until a fit between the unidentified specimen and an illustration, description, and range map is reached. Geography is a useful character, for if a certain species falls outside of your region, it may usually be eliminated from consideration, although of course we can expect range extensions as more collecting is done.

Not every specimen can be positively identified to species, because no key can possibly include all the variations of highly plastic genera with many "hybrids," such as *Bryoria*, *Cladonia*, *Collema*, *Ramalina*, and *Usnea*. A serious beginner is best advised to request help from lichenologists at universities or museums for identification or verification. Etiquette demands that only good, well labeled specimens be sent.

A millimeter scale is usually included with each illustration. Since enlargements vary, the scale should be noted with care. Lobe width is always given in millimeters whereas the diameter of the thallus is measured in centimeters. The color tests are indicated for most of the species along with the main chemical constituents. One should consult Culberson's book on lichen products for a complete inventory of the chemical composition of lichen species.

General References

A serious student will want to consult the references listed below, although some will be available only in university libraries. These articles treat the various species in more detail than is possible in this book and may have additional keys to forms and varieties. In particular, articles dealing with the flora of one state, as Ohio or Florida, will have shorter keys. The quarterly journal "The Bryologist" is the single most important reference for short scientific articles on lichens. One may subscribe to it for $15 per year. See a recent issue for subscription details.

AHMADJIAN, V. 1967. The lichen symbiosis. 152 pp. Blaisdell Publ. Co., Waltham, Mass. [Excellent summary of lichen physiology and synthesis; now out of print.]

––– and M. E. HALE. 1974. The Lichens. Academic Press, New York. [A text on most aspects of lichenology for advanced students.]

BOLTON, E. M. 1960. Lichens for vegetable dyeing. 119 pp. Charles T. Branford Co., Newton Centre, Mass. [Practical guide for lichen dyeing.]

BRODO, I. M. 1968. The lichens of Long Island, New York: a vegetational and floristic analysis. New York State Museum and Science Service, Albany, New York. [Detailed treatment useful for the Northeast, including keys to both crustose and macrolichens.]

––– and D. HAWKSWORTH. 1977. *Alectoria* and allied genera in North America. Opera Bot. 42 (Lund, Sweden). [Monograph of *Alectoria, Bryoria,* and *Pseudephebe*.]

CULBERSON, C. F. 1969. Chemical and botanical guide to lichen products. University of North Carolina Press, Chapel Hill. [Definitive treatment of lichen chemistry with an index of species and their chemistry.]

CULBERSON, W. L. and C. F. CULBERSON. 1968. The lichen genera *Cetrelia* and *Platismatia*. Contrib. U. S. Nat. Herb. 34:449-558. [Illustrated monograph including North American species.]

DEGELIUS, G. 1974. The lichen genus *Collema* with special reference to the extra-European species. Symbolae Bot. Upsalienses 20(2): 1-125 [A monograph including North American species.]

ESSLINGER, T. L. 1977. A chemosystematic revision of the brown *Parmeliae*. Jour. Hattori Bot. Lab. 42:1-211. [A world monograph with keys and illustrations.]

EVANS, A. W. 1943. Asahina's microchemical studies on the *Cladoniae*. Bull. Torrey Bot. Club 70:139-151. [Descriptions (but no illustrations) of crystals of the most common substances.]

FERRY, B. W., M. S. BADDELEY, and D. L. HAWKSWORTH. 1973. Air pollution and lichens. 389

pp. University of Toronto Press, Toronto. [Articles on all aspects of air pollution and lichens.]

FINK, B. 1935. The lichen flora of the United States. 426 pp. University of Michigan Press, Ann Arbor. [Complete summary with keys and descriptions but now out of date and out of print.]

HALE, M. E. 1954. Lichens from Baffin Island. Amer. Midl. Nat. 51:236-264. [Contains keys to many arctic lichens.]

———. 1961. Lichen handbook. 178 pp. Smithsonian Institution, Washington, D.C. [General brief summary of lichenology with keys to eastern lichens.]

———. 1974. The biology of lichens. 181 pp. Ed. Arnold (Publishers), London (available from American Elsevier Publ. Co., New York). [College-level text on lichenology.]

———. 1976a. A monograph of the lichen genus *Pseudoparmelia* Lynge (Parmeliaceae). Smithsonian Contr. Bot. 31:1-62. [A world monograph including North American species.]

———. 1976b. A monograph of the lichen genus *Parmelina* Hale (Parmeliaceae). Smithsonian Contr. Bot. 33:1-60. [A world monograph including North American species.]

——— and W. L. CULBERSON. 1970. A fourth checklist of the lichens of the Continental United States and Canada. Bryologist 73:499-543.

HENSSEN, A. 1963a. The North American species of *Placynthium*. Canadian Jour. Bot. 41:1688-1724.

———. 1963b. Eine Revision der Flechtenfamilien Lichinaceae und Ephebaceae. Symbolae Bot. Upsalienses 18(1):1-123. [A world monograph.]

IMSHAUG, H. A. 1957a. The lichen genus *Pyxine* in North and Middle America. Trans. Amer. Microsp. Soc. 76:246-269.

———. 1957b. Alpine lichens of western United States and adjacent Canada. I. The macrolichens. Bryologist 60:177-272. [Very useful treatment of alpine western lichens with complete keys but no illustrations.]

LLANO, G. A. 1950. A monograph of the lichen family Umbilicariaceae in the Western Hemisphere, 281 pp. Smithsonian Institution, Washington, D.C. [Available on request to university students.]

———. 1951. Economic uses of lichens. Ann. Rep. Smithsonian Institution 1950, pp. 385-422.

MOBERG, R. 1977. The lichen genus *Physcia* and allied genera in Fennoscandia. Symbolae Bot. Upsalienses 22(1):1-108. [Monographs of *Physcia*, *Physconia*, *Physciopsis*, and *Phaeophyscia*, including many species occurring in North America.]

MOORE, B. J. 1968. The macrolichen flora of Florida. Bryologist 71:161-265. [Complete with descriptions and keys and useful in general for the Coastal Plain from North Carolina to Texas.]

NEARING, G. G. 1947. The lichen book. [A complete but now out of date treatment of eastern lichens with numerous small line drawings.]

SHEARD, J. W. 1974. The genus *Dimelaena* in North America north of Mexico. Bryologist 77:128-141.

SIERK, H. A. 1964. The genus *Leptogium* in North America north of Mexico. Bryologist 67:245-317. [Monographic treatment with keys and illustrations.]

TAYLOR, FR. CONAN J. 1967. The foliose and fruticose lichens of Ohio. 100 pp. Ohio Biological Survey. [Excellent and well-illustrated flora of the Ohio lichens.]

THOMSON, J. W. 1950. The species of *Peltigera* of North America north of Mexico. Amer. Midl. Nat. 44:1-68. [Monographic treatment with keys and descriptions.]

———. 1963. The lichen genus *Physcia* in North America. Beih. Nova Hedwigia 7. 172 pp. [A monographic treatment with keys, descriptions, and photographs.]

———. 1967. The lichen genus *Cladonia* in North America. 200 pp. Univ. of Toronto Press, Toronto, Canada. [Complete illustrated treatment of *Cladonia*.]

WEBER, W. A. 1963. Lichens of the Chiricahua Mountains, Arizona. Univ. Colorado Studies. Biol. No. 10. [Includes a number of valuable keys to southwestern lichens.]

WETMORE, C. M. 1960. The lichen genus *Nephroma* in North and Middle America. Michigan State Univ. Biol. Series 1:372-452. [Monographic treatment with keys and descriptions.]

———. 1968. Lichen flora of the Black Hills. Michigan State Univ. Biol. series. [A complete survey with keys and descriptions for crustose and macrolichens; no illustrations.]

———. 1970. The lichen family Heppiaceae in North America. Annals Missouri Bot. Garden 57:158-209. [Complete summary of the family in North America with keys and illustrations.]

Pictured Key to the Foliose and Fruticose Lichens

1a Thallus foliose (see Fig. 3), more or less prostrate on the substratum, the lobes flattened or inflated with distinct upper and lower surfaces. Fig. 26. 2

Figure 26

Figure 26 Typical foliose lichen (*Parmotrema crinitum*)

1b Thallus umbilicate, fruticose, or squamulose. 3

2a (1) Thallus stratified (section with a razor blade as shown in Fig. 27), usually whitish mineral gray, yellowish green, orange, or brown with a white (rarely

pigmented) medulla. ....
.... (p. 27) I. *Stratified Foliose Lichens*

Figure 27

Figure 27 How to section a thallus with a razor blade to show stratified structure

2b Thallus without internal layers, usually dark bluish slate-colored to black; medulla dark. Fig. 28. ....
.... (p. 147) II. *Gelatinous Lichens*

Figure 28

Figure 28 Typical gelatinous lichen (*Collema subflaccidum*, ×1)

3a (1) Thallus umbilicate, round in outline and attached to rocks by a central cord below. Fig. 29. .... (p. 158) III. *Umbilicate Lichens*

Figure 29

Figure 29 Typical umbilicate lichens (*Umbilicaria vellea*)

3b   Thallus fruticose or squamulose. 4

4a (3) Thallus fruticose (see Fig. 10), free growing or attached at the base, the branches solid or hollow and round or flattened without distinct upper and lower surfaces. Fig. 30. .... (p. 167) IV. *Fruticose Lichens*

Figure 30

Figure 30 Typical fruticose lichens: *Usnea* sp. (left) and *Cladonia chlorophaea* (×1)

4b   Thallus squamulose (see Fig. 11), consisting of numerous separate squamules 1-10 mm long. Fig. 31. .... (p. 230) V. *Squamulose Lichens*

Figure 31

Figure 31 Squamulose growth form (*Cladonia*) (×2)

I. STRATIFIED FOLIOSE LICHENS

This is the largest and most frequently collected group of lichens. There is tremendous variation in thallus size and lobe width, but basically all species have more or less branched lobes, a distinct white (rarely pigmented) medulla, and a thin green (rarely blue-green) algal layer just below the upper cortex. They are also dorsiventral, that is there is a distinct upper surface and a lower surface that frequently, but not always, has a layer of rhizines. The species are separated by a number of characters, including thallus color, soredia, isidia, presence of rhizines or tomentum, cilia, etc.

The 44 genera included in this section belong to 13 different families, as tabulated in the phylogenetic listing (see page 239). While the genera are separated largely by spore characters and anatomy, it is difficult to construct a key to them since apothecia are lacking in many species. The Pictured Key uses thallus color as a main key character so as to group species of each genus as far as possible (as the brown Parmelias, yellow Xanthoparmelias, whitish gray Physcias, etc.). Of course when a genus has species of different color (as brown and yellow Cetrarias) they will be scattered throughout the keys. Lobe width, which should be measured with a ruler, not estimated, and color of the lower surface, black or brown to white, are also important key characters.

The following alphabetically arranged list gives brief salient features for each genus treated here and the numbers of the key couplets where the bulk of the species are keyed.

Acarospora (10): areolate crustose; apothecia immersed in areoles with numerous colorless simple spores per ascus.

Anaptychia (277, 280): foliose, narrow-lobed with or without a lower cortex; upper cortex prosoplectenchymatous; spores brown, 2-celled, 8/ascus.

Anzia (155): foliose, narrow-lobed with spongy black tomentum below; spores colorless, simple, many/ascus.

Bulbothrix (147, 172): foliose, mineral gray, narrow-lobed with inflated marginal cilia; spores colorless, simple, 8/ascus (formerly in *Parmelia*).

Caloplaca (4): subcrustose and lobate, orange (parietin); spores colorless, 2-celled and polarilocular, 8/ascus.

Candelaria (18): foliose, very narrow-lobed, yellow orange (calycin); spores colorless, simple, 8/ascus.

Candelina (11): subcrustose and lobate, yellow orange (calycin); spores colorless, simple, 8/ascus.

Cavernularia (167): foliose, narrow-lobed with numerous invaginations (pores) in the lower cortex; spores colorless, simple, 8/ascus.

Cetraria (19, 42, 228, 261, 273, 285): foliose or fruticose (flattened), narrow-lobed with mostly marginal erect pycnidia and apothecia; spores colorless, simple, 8/ascus.

Cetrelia (63): foliose, broad-lobed and pseudocyphellate; spores colorless, simple, 8/ascus.

Coccocarpia (143, 155): foliose, narrow-lobed with tomentum below; algae blue-green; spores colorless, simple, 8/ascus.

Dermatocarpon (257): subfoliose to umbilicate with perithecia.

Dimelaena (12): subcrustose and lobate; spores brown, 2-celled, 8/ascus (formerly in *Rinodina*).

Dirinaria (128, 132, 151, 170): foliose, narrow-lobed without rhizines; spores brown, 2-celled, 8/ascus; rim of apothecia lecanorine.

Everniastrum (126): foliose, narrow-lobed, canaliculate; spores colorless, simple, 8/ascus.

Heterodermia (179, 198): foliose, narrow-lobed, corticate or ecorticate below; upper cortex prosoplectenchymatous; spores brown, 2-celled, 8/ascus (formerly in *Anaptychia*).

Hypogymnia (101): foliose, narrow-lobed, hollow, lacking rhizines; spores colorless, simple, 8/ascus.

Hypotrachyna (126, 149, 174): foliose, narrow-lobed with dichotomously branched rhizines; spores colorless, simple, 8/ascus (formerly in *Parmelia*).

Lecanora (13): crustose and lobate; spores colorless, simple, 8/ascus.

Lobaria (39, 219, 253): foliose, mostly broad-lobed, tomentose below; algae green or blue-green; spores colorless, 1-9 septate, 8/ascus.

Menegazzia (100): foliose, narrow-lobed, hollow with perforations in upper cortex, lacking rhizines; spores colorless, simple, 8/ascus.

Nephroma (30, 226, 255, 263): foliose, medium- to broad-lobed, tomentose below; apothecia on lower surface of lobe tips; spores colorless, 2-4 septate, 8/ascus.

Neofuscelia (246, 286): foliose, brown, narrow-lobed, without pseudocyphellae; spores colorless, simple, 8/ascus (formerly in *Parmelia*).

Pannaria (233, 243, 278): foliose to squamulose, narrow-lobed, tomentose; algae blue-green; spores colorless, simple, 8/ascus.

Parmelia (16, 39, 62, 112, 142, 223, 238, 245): foliose, narrow- to medium-lobed, pseudocyphellate; spores colorless, simple, 8/ascus.

Parmeliella (279): foliose, narrow-lobed, tomentose below; algae blue-green; spores colorless, simple, 8/ascus.

Parmelina (119, 138, 142, 174): foliose, narrow-lobed, ciliate; spores colorless, simple, 8/ascus (formerly in *Parmelia*).

Parmeliopsis (22, 189, 198, 216): foliose, narrow-lobed; fulcra exobasidial; spores colorless, simple, 8/ascus.

Parmotrema (23, 70): foliose, broad-lobed, ciliate or eciliate; spores colorless, simple, 8/ascus (formerly in *Parmelia*).

Peltigera (46): foliose, broad-lobed, ecorticate and veined below; apothecia erect; spores colorless, 3-6 septate, 6-8/ascus.

Phaeophyscia (222, 231, 276, 289): foliose, narrow-lobed, atranorin lacking; lower cortex paraplectenchymatous, black; spores brown, 2-celled, 8/ascus (formerly in *Physcia*).

Physcia (177, 205): foliose, narrow-lobed, atranorin present; lower cortex usually prosoplectenchymatous, white; spores brown, 2-celled, 8/ascus.

Physciopsis (213): foliose, narrow-lobed, atranorin lacking; pale below; spores brown, 2-celled, 8/ascus (formerly in *Physcia*).

Physconia (224, 269, 282): foliose, narrow-lobed, often pruinose, dark below; spores brown, 2-celled, 8/ascus (formerly in *Physcia*).

Placopsis (132): subcrustose and lobate, with large cephalodia; spores colorless, simple, 8/ascus.

Platismatia (78, 91, 140, 158, 160): foliose, narrow- to broad-lobed, pseudocyphellate, lack-

ing rhizines; spores colorless, simple, 8/ascus.

Pseudevernia (140, 159): foliose, subfruticose, narrow-lobed, canaliculate, lacking rhizines; spores colorless, simple, 8/ascus.

Pseudocyphellaria (59): foliose, broad-lobed, tomentose and pseudocyphellate below; algae blue-green; spores brown, 1-3 septate, 8/ascus.

Pseudoparmelia (17, 44, 119, 195): foliose, narrow- to medium-lobed, eciliate; spores colorless, simple, 8/ascus (formerly in *Parmelia*).

Pyxine (111, 135): foliose, narrow-lobed, rhizinate and black below; spores brown, 2-celled, 8/ascus; rim of apothecia lecideine.

Solorina (252, 281): foliose, broad-lobed, tomentose below; algae blue-green; spores brown, septate, 2-8/ascus.

Sticta (56): foliose, broad-lobed, tomentose and cyphellate below; algae blue-green; spores colorless, 1-3 septate, 8/ascus.

Xanthoparmelia (27): foliose, narrow-lobed, yellow green (usnic acid), saxicolous, eciliate; spores colorless, simple, 8/ascus (formerly in *Parmelia*).

Xanthoria (3): foliose, orange, narrow-lobed; spores colorless, 2-celled and polarilocular, 8/ascus.

For convenience the main key groupings can be summarized as follows:

Orange lichens (*Caloplaca* and *Xanthoria*) — page 29
Yellowish green lichens — page 31
Veined lichens (*Peltigera*) — page 48
Pored lichens (*Cetrelia*, *Parmelia*, *Sticta*, and *Pseudocyphellaria*) — page 53
Broad-lobed mineral gray lichens (*Parmotrema* and *Platismatia*) — page 59
Narrow-lobed mineral gray lichens with a black lower surface — page 71
Narrow-lobed mineral gray lichens with a tan or white lower surface — page 101
Brown lichens — page 117

ORANGE LICHENS

1a   Thallus orange or yellowish to reddish orange, the surface (upper cortex) instantly K+ deep or blackish purple (parietin present). 2

1b   Thallus not orange; surface K−, K+ yellow or slowly greenish yellow (usnic acid, atranorin, brown pigments, or no substances present). 8

2a (1) Thallus growing on rocks. 3

2b   Thallus growing on trees. 5

3a (2) Thallus sorediate or isidiate. Fig. 32. ....
.... *Xanthoria sorediata* (Vain.) Poelt

Figure 32

Figure 32   Xanthoria sorediata

Thallus pale ochraceous orange, appressed, 2-4 cm broad but often fusing into large colonies; upper surface coarsely sorediate to sorediate-isidiate; lower surface white with sparse, coarse rhizines; apothecia lacking. Widespread on acidic rocks and overhanging ledges, especially along lake shores. *Xanthoria candelaria*,

which may also occur on rocks, is more distinctly foliose with soredia on the ends and lower side of the lobes.

3b   Thallus lacking soredia. 4

4a (3) Thallus closely attached but with a free lower surface and short rhizines (use hand lens). Fig. 33. ....
.... *Xanthoria elegans* (Link) Th. Fr.

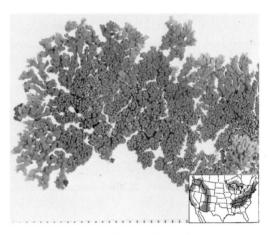

Figure 33

Figure 33   Xanthoria elegans

Thallus bright orange, closely adnate, 2-5 cm broad; lower surface white, with sparse, coarse rhizines; apothecia very common, crowded. Common on exposed cliffs and boulders. This is an extremely variable lichen which could be confused with subcrustose *Caloplaca saxicola*. When *X. parietina* is collected on seashore rocks it can be differentiated from *X. elegans* by the broader, more distinctly foliose lobes and more numerous rhizines.

4b   Thallus very closely attached to rock, crustose at the center, lacking rhizines. Fig. 34. ....
...... *Caloplaca saxicola* (Hoffm.) Nordin

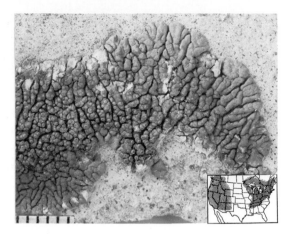

Figure 34

Figure 34   Caloplaca saxicola

Thallus dull yellowish orange to orange, often lightly white pruinose, closely adnate to sub-crustose, 1-3 cm broad, the lobes crowded, to 0.5 mm wide; apothecia numerous, crowded. Widespread on exposed limestone and other calcareous rocks. Along the coast of New England to Nova Scotia one will also find *C. scopularis* (Nyl.) Lett., similar in size and lobe configuration but bright orange and shiny and collected on acidic rocks rather than limestone. In the western states *C. saxicola* intergrades with *C. trachyphylla* (Tuck.) Zahlbr., a larger bright orange lichen with convex lobes to 1 mm wide.

5a (2) Thallus sorediate; apothecia rare. 6

5b   Thallus lacking soredia; apothecia usual-ly present. 7

6a (5) Lobes rather coarse, averaging 1 mm wide, little branched. Fig. 35. ....
.... *Xanthoria fallax* (Hepp) Arn.

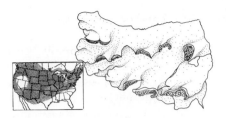

Figure 35

Figure 35   Xanthoria fallax (×5)

Thallus deep orange on exposed trees to pale greenish yellow on shaded trunks, closely ad-nate, 2-4 cm broad but often fusing into large colonies; lower surface white; apothecia rare. Widespread and commonly collected on oaks, aspen, and roadside elm trees. The lobes are quite coarse and unbranched when compared with those of typical *X. candelaria*, with which it intergrades, and the soralia are more labri-form and sometimes more or less swollen and hood-shaped.

6b   Lobes more finely divided, 0.2-0.5 mm wide. Fig. 36. ....
.... *Xanthoria candelaria* (L.) Th. Fr.

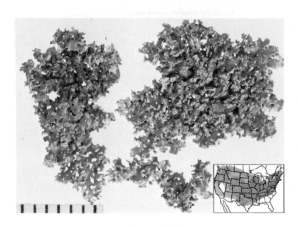

Figure 36

Figure 36 Xanthoria candelaria

Thallus orange to yellowish orange, closely adnate, fragile, 2-4 cm broad; soredia scattered but mostly apical; lower surface white; apothecia rare. Widespread on deciduous trees in open woods and along roadsides, more rarely on rocks. This species is more common than X. *fallax* in the western states and even occurs on the base of conifers. *Candelaria concolor* is very similar externally but has a yellow color and reacts K−.

7a (5) **Thallus small, 1-2 cm broad, the apothecia crowded. Fig. 37.** ....
.... *Xanthoria polycarpa* (Ehrh.) Oliv.

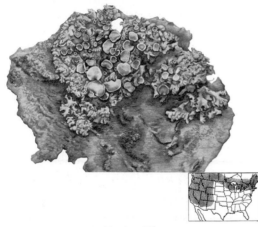

Figure 37

Figure 37 Xanthoria polycarpa (×5)

Thallus bright orange, closely adnate, the lobes often crowded and almost hidden by the numerous apothecia; lower surface white, sparsely rhizinate. Common on twigs, branches, and trunks of exposed trees, especially aspen. Specimens from the western states tend to be smaller than average; they have been called X. *ramulosa* (Tuck.) Herre but the taxonomy of the group is poorly known.

7b **Thallus larger and expanded, 3-6 cm broad, the apothecia numerous but not crowded. Frontispiece No. 1. Shore Lichen. .. Xanthoria parietina (L.) Th. Fr.**

Thallus deep to yellowish orange, shiny, adnate to loosely attached, the lobes often overlapping; lower surface ivory white with sparse, long rhizines. On trees and more rarely rocks near seashores. *Xanthoria elegans* can be differentiated by the narrower, more appressed lobes. Both species will be found on tombstones and stone fences.

YELLOWISH-GREEN LICHENS

8a (1) **Thallus (when dry) green, yellowish green, or pale to lemon yellow (usnic acid or yellow pigments present in the upper cortex; see Frontispiece Nos. 5, 6, 7).** 9

8b **Thallus mineral gray, white or brown (usnic acid absent; atranorin often present; see Frontispiece Nos. 8-13).** 45

9a (8) **Thallus crustose, firmly attached to rock at the center but usually becoming minutely lobed at the margins.** 10

9b **Thallus foliose with a lower surface and rhizines, closely adnate to loosely attached on bark, rock, or soil.** **14**

10a (9) Thallus entirely chinky, the margins not distinctly lobate. Frontispiece No. 2. .. ***Acarospora chlorophana* (Ach.) Mass.**

Thallus lemon yellow, crustose, forming discrete colonies 1-2 cm broad that may fuse to cover large areas, the margins irregular or becoming indistinctly lobate; apothecia visible only as pores in the areolae, the spores 32-64/ascus, simple and colorless. Common on exposed acidic rocks, especially in the western states. This is one of the most conspicuous crustose lichens, giving a yellow hue to whole mountainsides. The genus *Acarospora* contains many other yellow species but their taxonomy is not yet settled.

10b **Thallus chinky in the center but the margins distinctly lobate.** **11**

11a (10) Thallus brilliant yellow. Fig. 38. .... ***Candelina submexicana* (Lesd.) Poelt**

Figure 38

Figure 38   Candelina submexicana

Thallus closely appressed on rock, 1-3 cm broad, dull, the marginal lobes discrete, rather long, to 1 mm wide; apothecia common, the disk the same color as the thallus, the spores 8/ascus, simple and colorless. On acidic rocks in central and western Texas into New Mexico. A very similar species, *C. mexicana* (Lesd.) Poelt, has a brilliant yellow medulla and occurs with *C. submexicana*. The pigment in these two species is calycin, the same one known in *Candelaria concolor*.

11b **Thallus greenish yellow (usnic acid present).** **12**

12a (11) Spores brown, two-celled (use microscope). Fig. 39. .... ***Dimelaena oreina* (Ach.) Norm.**

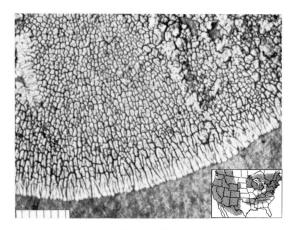

Figure 39

Figure 39   Dimelaena oreina

Thallus closely appressed, 2-6 cm broad, the marginal lobes distinct; apothecia very common, the disk black. Medulla K+ red (norstictic and/or stictic acid), K−, C+ red (gyrophoric acid), or K− C− (fumarprotocetraric acid if also P+ red or no acids present). Widespread on hard granitic rocks and quartzite in exposed areas. Other species in this genus are discussed by Sheard (1974) in a monograph.

12b   Spores colorless, one-celled (use micro-scope). **13**

13a (12) Lobe margins with a white or black rim, flattened. Fig. 40. .... *Lecanora muralis* **(Schreb.) Ach.**

Figure 40

Figure 40   Lecanora muralis

Thallus dull pale greenish yellow to whitish, closely appressed, 2-6 cm broad; lobes flat but with a somewhat raised white rim that blackens with age; apothecia numerous, the disk tan to brown. Medulla K−, C−, P− (leucotylin). Widespread on sandstone, granite, and calcareous rocks in open areas. This species is commonly found on stone walls and steps in farm yards.

13b   Lobes convex without any rim. Fig. 41. .. .. *Lecanora novomexicana* **Magn.**

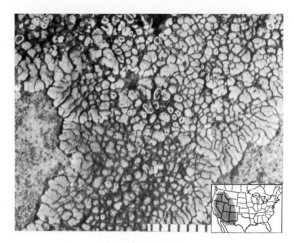

Figure 41

Figure 41   Lecanora novomexicana

Thallus deep greenish yellow, shiny, 1-2 cm broad but fusing to form large colonies, chinky at the center with long discrete marginal lobes to 1 mm wide, convex and often finely cracked, tightly appressed to the rock; apothecia common, the disk blackening. Common on sandstones and somewhat calcareous rocks. This is a robust and conspicuous lichen in the desert areas of the Southwest and at a distance can be mistaken for a *Xanthoparmelia,* which has a lower cortex and rhizines.

14a (9) Thallus sorediate along lobe margins or on the surface. **15**

14b   Thallus not sorediate. **23**

15a (14) Lobes quite broad and apically rotund, 3-10 mm wide (refer to Fig. 4 for measurement). **16**

15b   Lobes narrower, 0.1-3 mm wide, usually strap-shaped and linear. **18**

16a (15) Upper cortex with small white pores (look toward lobe tips with hand lens; also see Fig. 6B). Frontispiece No. 6 and Fig. 42. .... *Parmelia flaventior* Stirt.

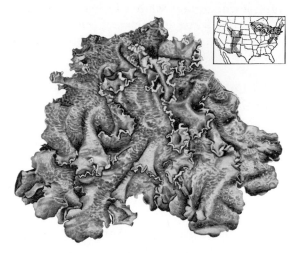

Figure 42

Figure 42   Parmelia flaventior (×1)

Thallus greenish yellow, loosely attached, 5-10 cm broad; soredia mostly laminal but also marginal; lower surface black with a broad naked brown zone along the margins; apothecia very rare. Medulla C+, KC+ red (lecanoric acid). Common on trees in open woods or along roadsides. This lichen has been confused with *Pseudoparmelia caperata*, which differs in lacking pores and in having a C−, P+ red reaction in the medulla as well as laminal, rather diffuse soralia. *Parmelia ulophyllodes* (below) is very close but has few or no pores.

16b   Upper cortex continuous, without white pores. 17

**17a (16) Soredia on the surface of lobes, often diffuse and granular. Frontispiece No. 7 and Fig. 43. ....
...... *Pseudoparmelia caperata* (L.) Hale**

Figure 43

Figure 43   Pseudoparmelia caperata (×1)

Thallus pale greenish yellow, adnate to loosely attached, 5-20 cm broad, often covering large areas of tree trunks; upper surface smooth to wrinkled; lower surface black and rhizinate with a narrow brown bare zone at the margin; apothecia very rare. Medulla K−, P+ red (protocetraric acid). Very common on trees in open woods and along roads, very rarely on rocks (except in northern part of range). *Parmelia flaventior* is superficially similar but has pores and reacts C+ red. Rock-inhabiting *Pseudoparmelia baltimorensis* (see p. 38) is very close but has isidia-like pustules without soredia.

**17b   Soredia along margins of lobes in linear soralia. Fig. 44. ....
.... *Parmelia ulophyllodes* (Vain.) Sav.**

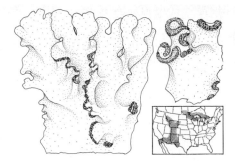

Figure 44

Figure 44   Parmelia ulophyllodes (×3)

Thallus yellowish green, adnate, 4-8 cm broad; upper surface rather wrinkled; sorediate lobes in part suberect; lower surface black or dark brown, naked and lighter brown near the margins; apothecia very rare. Medulla C+, KC+ red (lecanoric acid). On conifers and deciduous trees in open woods. This species intergrades with *P. flaventior* extensively in western states and it may be difficult to make firm determinations.

18a (15) Lobes finely branched, 0.1-0.3 mm wide, soredia scattered. Fig. 45. ....
...... *Candelaria concolor* (Dicks.) Stein.

Figure 45

Figure 45   Candelaria concolor

Thallus greenish lemon yellow, closely adnate, 0.2-1.0 cm broad but fusing into larger colonies; lower surface white, rhizinate; apothecia rare. Common on ash, elm, and sugar maple in open woods. It reacts K− in contrast to the K+ purple of *Xanthoria candelaria*.

18b   Lobes broader, often little branched, 0.5-3 mm wide. 19

19a (18) Medulla and soredia deep lemon yellow. Fig. 46. Pine Lichen. ....
.... *Cetraria pinastri* (Scop.) S. Gray

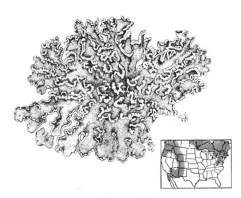

Figure 46

Figure 46   Cetraria pinastri (×2)

Thallus light yellow green, adnate, 1-2 cm broad; lower surface white, sparsely rhizinate; apothecia very rare. Common and conspicuous at the base of conifers in boreal areas and the Rocky Mountains. This brilliant lichen often occurs with *Parmeliopsis ambigua*. The lemon yellow pigment is pinastric acid.

19b   Medulla and soredia white. 20

20a (19) Soredia mostly on margins of lobes in linear soralia. Fig. 47. ....
.... *Cetraria oakesiana* **Tuck.**

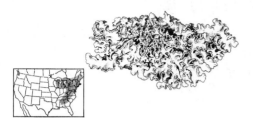

Figure 47

Figure 47   Cetraria oakesiana (×2)

Thallus yellowish green, adnate, 3-7 cm broad, the lobes often more or less parallel; lower surface light tan to white, sparsely rhizinate; apothecia rare; erect black marginal pycnidia sometimes present. Common on the bark of conifers and hardwoods and on rocks in northern woods. Forms of *Parmelia ulophyllodes* with a pale lower surface may key out here, but they have broader lobes and a C+ red reaction in the medulla. This species is C− and contains fatty acids.

20b   Soredia on surface or tips of lobes in orbicular soralia. **21**

21a (20) Medulla K+ yellow turning red; rhizines dichotomously branched (use hand lens). Fig. 48. ....
.... *Hypotrachyna sinuosa* **(Sm.) Hale**

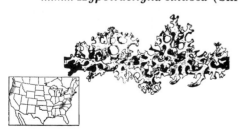

Figure 48

Figure 48   Hypotrachyna sinuosa (×1)

Thallus greenish yellow, loosely adnate to somewhat suberect on twigs, 2-6 cm broad; apothecia lacking. Medulla K+ red, P+ orange (salazinic acid). Rarely collected on trees in open woods in the Cascades of Oregon and Washington and at high elevation in the southern Appalachians. The species was previously classified in *Parmelia*. A rare rock-inhabiting *Xanthoparmelia* in Colorado, *X. mougeotii* (Pers.) Hale, will key here because of the K+ test (usnic and stictic acids), although the rhizines are simple.

21b   Medulla K−; rhizines simple. **22**

22a (21) Collected in northern and western states and Canada. Fig. 49. ....
...... *Parmeliopsis ambigua* **(Wulf.) Nyl.**

Figure 49

Figure 49   Parmeliopsis ambigua

Thallus light greenish yellow, closely adnate on bark, 2-4 cm broad; lobes narrow and linear, about 1 mm wide; soredia in powdery laminal soralia; lower surface dark brown to blackish; apothecia not common. Medulla K−, C− (usnic and divaricatic acids). Widespread at the base of conifers and deciduous trees and on bare wood, stumps, and logs. A very similar species, *P. capitata* R. Harris, has a white to tan lower surface and soralia mostly subterminal; it occurs in the Great Lakes-Appalachian region. On the north shore of Lake Superior and similar alpine habitats in the western states one may find *Xanthoparmelia subcentrifuga* (Oxner) Hale; it is a saxicolous lichen with broader lobes than *P. ambigua*, coarse pustular soredia, and KC+ red medulla (alectoronic acid). The lower surface is black.

22b   Collected in the Piedmont and Coastal Plain of southeastern United States. Fig. 50. .... *Parmeliopsis subambigua* Gyel.

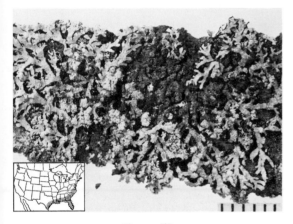

Figure 50

Figure 50   Parmeliopsis subambigua

Thallus greenish yellow, closely adnate, 2-4 cm broad; upper surface pustulate and diffusely sorediate; lower surface uniformly whitish to cream; apothecia rare. Common on pine trees and burned stumps in open woods and along roadsides. This species was formerly identified with *P. ambigua*.

23a (14) Thallus isidiate (see Fig. 51). 24

Figure 51

Figure 51   Cylindrical isidia of *Parmotrema xanthinum* (A), pustulate isidia of *Pseudoparmelia baltimorensis* (B), and diffuse soralia of *P. caperata* (C) (all about ×10)

23b   Thallus lacking isidia. 29

24a (23) Medulla brilliant yellow (expose with razor blade). Fig. 51. .... *Parmotrema sulphuratum* (Nees & Flot.) Hale

Figure 52

Figure 52   Parmotrema sulphuratum

Thallus yellowish green, adnate, 6-10 cm broad; upper cortex becoming cracked with age, exposing the yellow medulla; lobe margins becoming dissected, isidiate, sparsely ciliate; lower surface with a narrow bare zone along the margins; apothecia lacking. Cortex K+ yellow (atranorin); vulpinic acid in the medulla. Rather rare on hardwoods in mature forests. The most commonly collected *Parmotrema* with a yellow medulla will probably be *P. endosulphureum* (see p. 65), which has a paler orange-yellow medulla and lacks cilia.

24b **Medulla white.** 25

25a **(24) Isidia coarse, to 1 mm in diameter, breaking open with age (see Fig. 52). Fig. 53.** .... *Pseudoparmelia baltimorensis* **(Gyel. & For.) Hale**

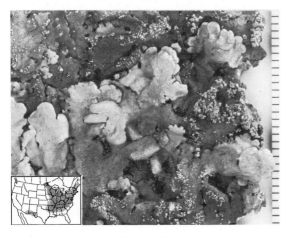

Figure 53

Figure 53 Pseudoparmelia baltimorensis

Thallus yellowish green, adnate to loosely attached on rock, 6-15 cm broad with colonies fusing to cover large areas; upper cortex becoming rugose with age and covered with coarse isidia which may appear sorediate; lower surface black with a narrow bare brown zone at the margins; apothecia rare. Medulla K−, C+ red or C−, P+ red (protocetraric acid with or without gyrophoric acid). Very common on rocks in open oak woods. This lichen

had been previously confused with *Pseudoparmelia caperata* which differs in having coarse soredia and in almost always growing on trees.

25b **Isidia fine and cylindrical, not more than 0.3 mm wide and not breaking open (use hand lens and see Fig. 52).** 26

26a **(25) Lobes 3-10 mm broad, marginally ciliate. Fig. 54.** .... *Parmotrema xanthinum* **(Müll. Arg.) Hale**

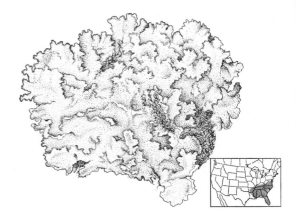

Figure 54

Figure 54 Parmotrema xanthinum (×1)

Thallus yellowish green, loosely adnate, 8-20 cm broad; isidia simple to coralloid; lobe margins becoming dissected and isidiate; lower surface with a broad bare marginal zone; apothecia lacking. Medulla K−, C−, P− (fatty acids). Rather rare on tree trunks and boulders in open woods. A chemical variant with gyrophoric acid (medulla C+ rose) is called *P. madagascariaceum* (Hue) Hale and occurs with *P. xanthinum*.

26b **Lobes narrower, 1-4 mm broad, lacking cilia.** 27

27a (26) Medulla K— (see Fig. 18 for sectioning technique). Fig. 55. .... *Xanthoparmelia subramigera* (Gyel.) Hale

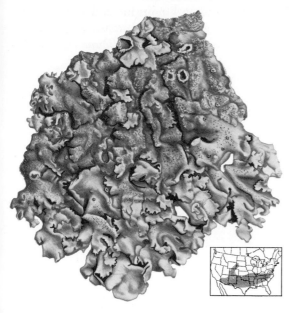

Figure 55

Figure 55 Xanthoparmelia subramigera (×1)

Thallus yellowish green, adnate, 4-8 cm broad, the lobes short and sometimes crowded; isidia quite dense, mostly simple; lower surface uniformly tan, moderately rhizinate. Medulla K—, C—, P+ red (fumarprotocetraric and succinprotocetraric acids). Widespread on exposed rocks in arid regions. There are a number of isidiate Xanthoparmelias in southwestern United States with a pale lower surface but they can only be identified with chemical tests since they are more or less indistinguishable externally. *Xanthoparmelia weberi* (Hale) Hale, for example, contains hypoprotocetraric acid (medulla K—, C—, P—) and is very common in Arizona and New Mexico. Rare *X. joranadia* (Nash) Hale reacts C+ red (lecanoric acid), while *X.*

kurokawae (Hale) Hale is P+ yellow (psoromic acid) and *X. ajoensis* (Nash) Egan is negative (diffractaic acid).

27b Medulla K+ yellow or yellow turning red. 28

28a (27) Lower surface more or less uniformly black. Fig. 56. Boulder Lichen. .... *Xanthoparmelia conspersa* (Ach.) Hale

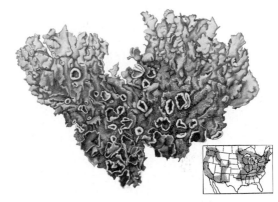

Figure 56

Figure 56 Xanthoparmelia conspersa (×1)

Thallus yellowish green, usually rather adnate (loosely adnate in some specimens from southern United States), 4-12 cm broad, colonies often fusing to cover very large areas of rock; isidia sparse to dense, simple; lower surface moderately rhizinate; apothecia common. Medulla K+ yellow, P+ orange (stictic and norstictic acids). Extremely common on exposed granite and sandstone throughout North America. Most specimens in eastern North America contain stictic acid, but there are other chemical variants that can be identified only with crystal or color tests: *X. piedmontensis* Hale (P+ red, K—, fumarprotocetraric acid), rare in the southern Appalachians, *X. tinctina* (Mah. & Gill.) Hale (K+ red, salazinic acid), rather rare in the north central and western states, and *X. congensis* (Stein.) Hale (K+ yellow, stictic acid and norstictic acid), a rare lichen in the southern states with very narrow lobes, less than 1 mm wide.

28b **Lower surface uniformly tan to light brown. Fig. 57.** .... ***Xanthoparmelia plittii*** (**Gyel.**) **Hale**

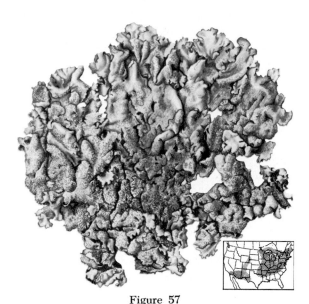

Figure 57

Figure 57 Xanthoparmelia plittii (×1)

Thallus, chemistry, and distribution as in *X. conspersa* but the lower surface brown. Very common on exposed acidic rocks. Chemical variants, which occur in the western states, include *X. mexicana* (Gyel.) Hale (medulla K+ red, salazinic acid), common from Montana and Minnesota southward, and rare *X. dierythra* (Hale) Hale (medulla K+ red, norstictic acid only).

29a (23) Collected on rocks, soil, or humus. **30**

29b **Collected on trees.** **37**

30a (29) Lobes very broad, 20-30 mm across; collected on humus in arctic-alpine habitats. Fig. 58. .... ***Nephroma arcticum*** (**L.**) **Torss.**

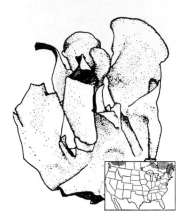

Figure 58

Figure 58 Nephroma arcticum (×1)

Thallus yellowish green, loosely attached on soil and among mosses, 6-15 cm broad; lower surface black and short tomentose at the center, tan toward the margin; apothecia common, **up** to 2 cm in diameter, on the lower surface of lobe tips. Widespread and conspicuous. This is one of the most unusual arctic lichens.

30b **Lobes narrower, 0.5-5 mm wide; collected on rocks or soil mostly in temperate zones.** **31**

31a (30) Collected growing loose on soil. Fig. 59. .... ***Xanthoparmelia chlorochroa*** (**Tuck.**) **Hale**

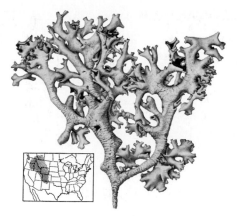

Figure 59

Figure 59   Xanthoparmelia chlorochroa (×1)

Thallus greenish yellow, leathery, composed of separate groups of scattered lobes, the margins turning under, 2-3 cm long; lower surface tan to dark brown, sparsely rhizinate; apothecia very rare. Medulla K+ yellow→red, P+ orange (salazinic acid or rarely stictic acid). Locally abundant growing loose on soil among prairie grasses. This distinctive lichen intergrades with western forms of *X. taractica* (see below). At higher elevations in the Rocky Mountains it intergrades with *X. wyomingica* (Gyel.) Hale, which also has salazinic acid and convoluted lobes but is a smaller lichen with lobes only 1-2 mm wide.

31b   Collected on rocks. 32

32a (31) Lower surface uniformly white to tan or pale brown. 33

32b   Lower surface black with dark brown margins. 36

33a (32) Lower surface white; collected in arctic-alpine habitats. Fig. 60. ....
.... *Xanthoparmelia centrifuga* (L.) Hale

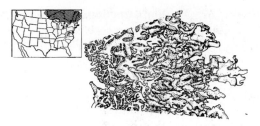

Figure 60

Figure 60   Xanthoparmelia centrifuga (×1)

Thallus pale greenish yellow, closely adnate, often forming concentric bands, 3-10 cm broad; lower surface sparsely rhizinate; apothecia rare. Medulla K−, KC+ red, P− (alectoronic acid). Common on rocks in arctic regions, rarer southward along the northern Great Lakes shore line and high elevations in the Appalachians.

33b   Lower surface tan to pale brown; collected in temperate or less commonly alpine habitats. 34

34a (33) Medulla K− (see Fig. 18 for sectioning technique). Fig. 61. .... *Xanthoparmelia novomexicana* (Gyel.) Hale

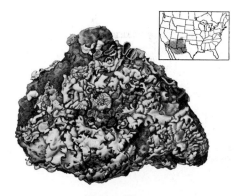

Figure 61

Figure 61   Xanthoparmelia novomexicana

Thallus yellowish green, closely adnate, the lobes rather narrow, often no more than 1 mm wide, forming small colonies 3-6 cm broad; lower surface tan, moderately rhizinate; apothecia common. Medulla K−, C−, P+ red (fumarprotocetraric acid). Widespread on rocks in arid regions. Several unrelated and rather rare Xanthoparmelias also react K− and have a pale brown lower surface and will have to be separated with chemical tests. In the western states there are three species, all with larger thalli and lobes 1-3 mm wide: *X. subdecipiens* (Vain.) Hale (medulla K−, C−, P−, fatty acids), *X. psoromifera* (Kurok.) Hale (medulla K−, C−, P+ yellow, psoromic acid), and *X. tucsonensis* (Nash) Egan (medulla K−, C−, P−, diffractaic acid). The only eastern representative, rare *X. monticola* (Dey) Hale in the southern Appalachians, is more loosely attached and has the same chemistry as *X. novomexicana* in addition to stictic acid.

34b   Medulla K+ yellow or yellow turning red. **35**

35a (34)   Thallus rather tightly attached to rocks and not easily removed with a knife; lobes rather short and crowded. Frontispiece No. 5 and Fig. 62. .... *Xanthoparmelia cumberlandia* (Gyel.) Hale

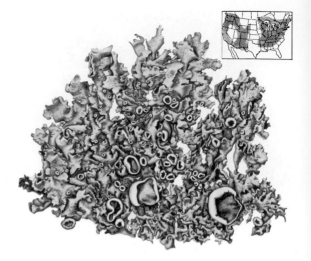

Figure 62

Figure 62   Xanthoparmelia   cumberlandia (×1.5)

Thallus greenish yellow, forming orbicular colonies 3-12 cm broad, the central part tending to become densely lobulate with age in some specimens; lower surface tan and moderately rhizinate; apothecia numerous. Medulla K+ yellow or yellow turning red, C−, P+ orange (stictic and norstictic acids). Common on exposed rocks. A narrow-lobed, tightly appressed variant in the southwestern deserts is known as *X. arseneana* (Gyel.) Hale but the two species intergrade. Two other species which differ in containing salazinic acid (medulla K+ finally red) are common in the western states. *Xanthoparmelia lineola* (Berry) Hale is very similar to *X. cumberlandia* externally, and *X. ioannis-simae* (Gyel.) Hale has broader lobes (to 5 mm wide) with rounded tips. There are, however, a number of taxonomic problems with these species.

35b   Thallus loosely attached on rock and easily removed with a knife; lobes generally long and linear. Fig. 63. .... *Xanthoparmelia taractica* (Kremplh.) Hale

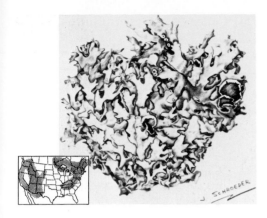

Figure 63

Figure 63 Xanthoparmelia taractica (×1)

Thallus greenish yellow, 4-15 cm broad; lobes quite elongate and linear; lower surface tan, moderately rhizinate; apothecia common. Medulla K+ yellow→red, P+ orange (salazinic acid). Widespread on exposed acidic rocks but not as common as *X. cumberlandia*. This is a variable species that intergrades with both *X. chlorochroa*, which is free on soil, and with *X. lineola*, which is more adnate. Specimens with a black lower surface are *X. tasmanica* (see below).

36a (32) Thallus tightly attached to rock and not easily removed with a knife; lobes rather short and crowded. Fig. 64. .. *Xanthoparmelia hypopsila* (Müll. Arg.) Hale

Figure 64

Figure 64 Xanthoparmelia hypopsila (×1.5)

Thallus greenish yellow, rather closely attached to rock, the lobes crowded, forming colonies 4-8 cm broad; lower surface black and moderately rhizinate; apothecia common. Medulla K+ yellow or yellow turning red, C−, P+ orange (stictic and norstictic acids). On exposed rocks. This lichen is much rarer than its close relative, *X. cumberlandia,* which is pale below. There are a number of exclusively western species which will key here because of the close attachment and black lower surface but which differ principally in chemistry. All are K negative. *Xanthoparmelia dissensa* (Nash) Hale, a rare species in Arizona and New Mexico, contains hypoprotocetraric acid (P−); *X. huachucensis* (Nash) Egan, another Arizona species, is P+ yellow (psoromic acid). At this time there is only one member of the group in eastern North America, *X. hypomelaena* (Hale) Hale (medulla P+ red, fumarprotocetraric acid), which occurs in Arkansas and neighboring states.

36b Thallus loosely attached on rock and easily removed with a knife; lobes generally narrow and long. Fig. 65. .... *Xanthoparmelia tasmanica* (Hook. & Tayl.) Hale

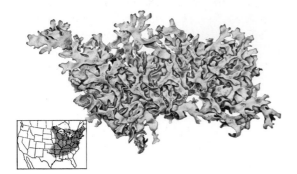

Figure 65

Figure 65   Xanthoparmelia tasmanica

Thallus, chemistry, and distribution as in *X. taractica* but the lower surface black, turning brown only at the margins. This species appears to be quite common and often occurs with *X. taractica.*

37a (29) Collected in western North America. 38

37b Collected in eastern and southern United States. 41

38a (37) Thallus large with broad lobes 10-30 mm wide. 39

38b Thallus smaller with lobes 1-5 mm wide. 40

39a (38) Upper surface heavily reticulately ridged (without lens); pores lacking. Fig. 66. Horse Lettuce. .... *Lobaria oregana* (Müll. Arg.) Hale

Figure 66   Lobaria oregana (×1)

Thallus light greenish yellow, loosely attached, 10-25 cm broad; lower surface mottled brown and cream, short tomentose; apothecia rare. Medulla K+ yellow, P+ orange (stictic acid). Common on conifers in open woods or on mos-

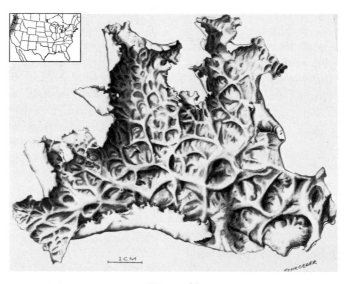

Figure 66

sy rocks. The ridged upper surface is similar to that of *Lobaria linita* and *L. pulmonaria* but the color and chemistry differ. This unusual lichen is especially common in the Cascades of Oregon, where one will often find large clumps of thalli that have fallen to the forest floor.

39b Upper surface smooth, not ridged; white pores present (as in Fig. 6B). Fig. 67. ....
.... *Parmelia praesignis* Nyl.

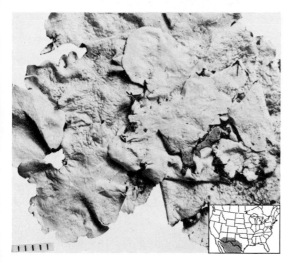

Figure 67

Figure 67 Parmelia praesignis

Thallus greenish yellow, adnate, 6-15 cm broad; upper surface with white pores (pseudocyphellae); lower surface black and sparsely rhizinate; apothecia common. Medulla C+, KC+ red (lecanoric acid). Common on tree trunks in open woods. This conspicuous lichen is the nonsorediate form of the common *P. flaventior* (see p. 34). In southern Arizona one may collect *P. darrovii* Thoms., a rare variant with a uniformly tan lower surface.

40a (38) Medulla sulfur yellow (expose with razor blade). Frontispiece No. 3. ....
.... *Cetraria canadensis* Räs.

Thallus loosely attached to bark, adnate to suberect, forming almost orbicular colonies 2-4 cm broad; upper surface ridged and wrinkled (with hand lens); lower surface wrinkled, very sparsely rhizinate; apothecia very common, marginal, the disc dark brown. Medulla K+ yellowish, C−, P− (pinastric and vulpinic acids). Very common on conifers in open forests. A few isolated specimens will produce soredia. This is one of the most common and conspicuous lichens in the northern Rocky Mountains and northern California, often occurring with brown *C. platyphylla*. Another suberect lichen in the Cascades, *C. pallidula* Tuck., is pale yellowish green and has a white medulla.

40b Medulla white. Fig. 68. ....
...... *Parmelia sphaerosporella* Müll. Arg.

Figure 68

Figure 68 Parmelia sphaerosporella

Thallus light yellowish green, closely adnate, 4-8 cm broad; upper surface finely wrinkled (without lens); lower surface buff to white, moderately rhizinate; apothecia common. Widespread on conifers, especially *Abies* and *Picea*, at higher elevations.

41a (37) Thallus lemon yellow; lobes finely divided, 0.3-0.5 mm wide. Fig. 69. .... *Candelaria fibrosa* (Fr.) Müll. Arg.

Figure 69

Figure 69   Candelaria fibrosa

Thallus greenish lemon yellow, 1-2 cm broad, closely adnate; lower surface white, sparsely rhizinate; apothecia numerous, with rhizines around the base. Cortex K+ yellowish (calycin). Widespread on deciduous trees in open woods or along roadsides. This is the nonsorediate counterpart of *Candelaria concolor* (see p. 35), which is much more common.

41b   Lobes broader, 2-5 mm wide. 42

42a (41) Medulla sulfur yellow (expose with razor blade). Fig. 70. .... *Cetraria viridis* Schwein.

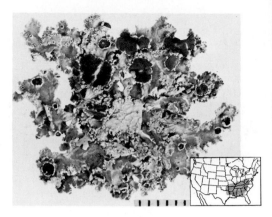

Figure 70

Figure 70   Cetraria viridis

Thallus light yellowish green adnate, 2-3 cm broad; lower surface light yellow, sparsely rhizinate; apothecia common. Vulpinic acid in the medulla. Widespread but rather rare on branches of deciduous trees in open woods. This species often grows in the canopies of trees.

42b   Medulla white (except turning pale orange in *Pseudoparmelia sphaerospora*). 43

43a (42) Collected on pine trees in the Appalachian-Great Lakes region. Fig. 71. .... *Cetraria aurescens* Tuck.

Figure 71

Figure 71   Cetraria aurescens (×3)

Thallus light yellowish green, adnate, 2-6 cm broad; upper surface somewhat ridged (without lens); lower surface light tan, sparsely rhizinate; apothecia and pycnidia common. Common on trunks and branches of pine trees in open woodlots. It is often collected with *Parmeliopsis placorodia*.

43b   Collected in southern United States (into Texas). 44

44a (43) Lower surface black. Fig. 72. .... *Pseudoparmelia rutidota* (**Hook. & Tayl.**) **Hale**

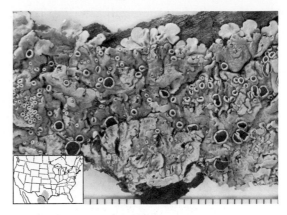

Figure 72

Figure 72   Pseudoparmelia rutidota

Thallus yellowish green, closely adnate on bark, 4-8 cm broad; lower surface moderately rhizinate with a narrow bare zone at the margin; apothecia very common, the disk dark brown. Medulla K−, C−, P+ red (protocetraric acid). On trees in open areas, mostly in Texas. This species is close to *P. caperata*, which very rarely lacks soredia in the juvenile stages and would key out here. Such specimens; however, would probably lack apothecia and be collected in the eastern states.

44b Lower surface uniformly tan. Fig. 73. ....
.... *Pseudoparmelia*
sphaerospora (Nyl.) Hale

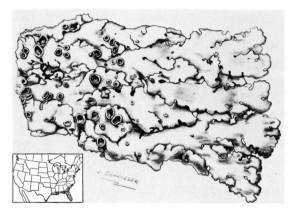

Figure 73

Figure 73 Pseudoparmelia sphaerospora (×2)

Thallus pale yellowish mineral gray, closely adnate, 5-10 cm broad; medulla white to pale yellow orange; lower surface pale buff, moderately rhizinate; apothecia very common. Medulla K+, C+, P+ yellowish (stictic acid and unidentified substances). Common on deciduous trees in humid forests. This species will be found in the range of Spanish Moss in the southern states.

45a (8) Lower surface fibrous or cottony (use hand lens), with raised pale veins or flat dark veins (hand lens not needed; see Figs. 74 and 507); lobes quite broad, usually 10 mm or more wide; thallus often growing on soil, among mosses, or on humus over rocks, more rarely on tree bark. 46

Figure 74

Figure 74 Lower surface of *Peltigera* species showing types of veins (×1)

45b Lower surface without any veins (hand lens not needed), smooth or wrinkled, bare, tomentose, or rhizinate (ecorticate and cottony in the narrow-lobed genus *Heterodermia* and brown-white mottled in *Lobaria* mimicking veins); thallus with broad or narrow lobes, collected most often on trees and rocks, more rarely on soil. 54

VEINED LICHENS (PELTIGERA)

46a (45) Upper surface of lobes with small scattered dark green warts (cephalodia) (without lens); thallus turning bright green when moist. Fig. 75. ....
.... *Peltigera aphthosa* (L.) Willd.

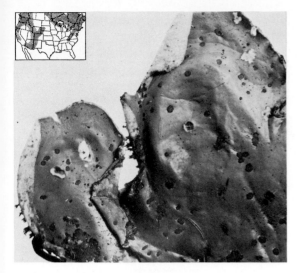

Figure 75

Figure 75 Peltigera aphthosa (×2)

Thallus pale greenish gray (when dry), loosely adnate to suberect on soil and over mosses, forming colonies to 1 mm wide; lower surface dirty whitish, the veins rather pale and indistinct; apothecia common, erect. Common and conspicuous in mountainous areas. The warty cephalodia (see also Fig. 6G) are distinctive. In the eastern part of North America most of the specimens have dark, distinct veins and may be identified as *P. leucophlebia* (Nyl.) Gyel.

46b Warts lacking on the upper surface; thallus not turning bright green when wet (except in *P. malacea* and *P. venosa*). ....
.... 47

47a (46) Thallus with laminal or marginal soredia. 48

47b Thallus lacking soredia. 49

48a (47) Soredia laminal in orbicular soralia. Fig. 76. ....
.... *Peltigera spuria* (Ach.) DC.

Figure 76

Figure 76 Peltigera spuria

Thallus brownish gray, adnate, 3-7 cm broad; upper surface finely tomentose; lower surface light tan, the veins raised, pale or brownish, rhizines distinct; apothecia very rare. Scattered on soil in woods or along roadbanks but often overlooked. This species is quite evanescent and colonies can disappear in a year. It is related to *P. canina* and may only be a transitional growth form.

48b Soredia in marginal, linear soralia. Fig. 77. .... *Peltigera collina* (Ach.) Ach.

Figure 77

Figure 77   Peltigera collina (×1.5)

Thallus greenish or brownish gray, 5-10 cm broad; upper surface becoming scabrid near the tips; margins curled upward, densely sorediate; lower surface tan, the veins flattened and indistinct, darkening, rhizines sparse; apothecia not common, erect. Common on soil, mosses, humus over boulders, and base of trees. It is not easily confused with any other species. The medulla below is sometimes very dense and the veins not easily distinguished.

49a (47) Thallus isidiate. Fig. 78. ....
.... *Peltigera evansiana* Gyel.

Figure 78

Figure 78   Peltigera evansiana

Thallus light brown, adnate, 6-12, cm broad; isidia short and globular; lower surface pale, the veins distinct, rhizines usually well developed; apothecia very rare. Common on soil in fairly closed woods and on sheltered roadbanks. The isidia distinguish this rather rare species from *P. canina*. In western North America, especially in Colorado north to Alberta, rarely eastward, one will find another isidiate species, *P. lepidophora* (Nyl.) Vain., which has much coarser, peltate isidia.

49b   Thallus lacking isidia (flattened tiny lobules sometimes present). 50

50a (49) Lower surface with pale, narrow, raised veins and long, conspicuous rhizines. Fig. 79. Dog Lichen. ....
.... *Peltigera canina* (L.) Willd.

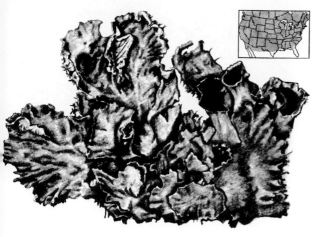

Figure 79

Figure 80

Figure 79   Peltigera canina (×1)

Thallus light brown with a whitish cast when dry, deeper brown when wet, adnate on soil, humus, or mosses, 6-20 cm broad; upper surface dull with a fine tomentum, especially toward the tips (Fig. 80); lower surface tan to whitish, without a cortex, the veins raised, the same color or darkening, rhizines pale, tufted and conspicuous; apothecia common, erect. Extremely widespread throughout most of North America. This lichen is highly variable. A darker brown sun-form with smaller curled lobes is sometimes recognized as *P. rufescens* (Weiss) Humb. Another species known from central United States, *P. degenii* Gyel., has an identical lower surface with distinct raised veins but the upper surface is shiny and lacks tomentum. When the thallus of *P. canina* breaks or is torn, small lobules may regenerate along the edges. This variant may be recognized as a distinct species, *P. praetextata* (Somm.) Vain. Another species in this highly plastic group, *P. membranacea* (Ach.) Nyl., has very broad lobes (4-6 cm wide) and a very thin thallus. It appears to be most common in the Pacific Northwest. If the lower surface is brilliant orange, you have collected *Solorina crocea*.

Figure 80   Tomentum on the surface of *Peltigera canina* (×15)

50b   Lower surface with dark, flattened veins or veins not clearly distinguishable; rhizines short or lacking. 51

51a (50)   Thallus small, 1-2 cm broad, fan-shaped. Fig. 81. Fan Lichen. ....
.... *Peltigera venosa* (L.) Baumg.

Figure 81

Figure 81   Peltigera venosa (×3)

Thallus dull green to light brown, adnate to suberect on soil, composed of separate lobes; upper surface shiny and smooth; lower surface with dark veins sparsely covered with tiny greenish warts (cephalodia); rhizines inconspicuous; apothecia common, horizontal. Rather rare except in boreal areas and easily overlooked because of the small size. A frequent habitat is poorly consolidated soil banks along trails and roads.

51b Thallus larger, 6-20 cm broad with long lobes. **52**

52a (51) Apothecia always present, oriented horizontally. Fig. 82. .... *.. Peltigera horizontalis* **(Huds.) Baumg.**

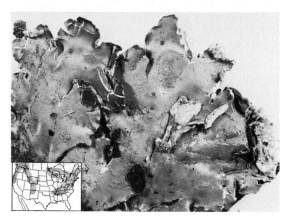

Figure 82

Figure 82   Peltigera horizontalis (×2)

Thallus brownish mineral gray, loosely adnate, 6-12 cm broad; upper surface shiny; lower surface buff, the veins flattened, rhizines sparse. Fairly common in moist woods. This species is separated from *P. polydactyla* chiefly by the horizontal apothecia. Sterile specimens of *P. horizontalis* would therefore be identified as *P. polydactyla*.

52b Apothecia lacking or, if present, erect. **53**

53a (52) Upper surface shiny (use hand lens). Fig. 83. .... *.. Peltigera polydactyla* **(Neck.) Hoffm.**

Figure 83

Figure 83   Peltigera polydactyla (×1)

Thallus brownish gray, the lobes broad, semierect, forming extensive colonies; lower surface buff, the veins darkening (Fig. 74); apothecia common, erect. Very common on soil and over mosses in open forests. Some forms develop numerous tiny lobules along cracks; these may be recognized as *P. elizabethae* Gyel., an eastern species.

53b Upper surface dull and scabrid (Fig. 80). Fig. 84. .... *Peltigera malacea* **(Ach.) Funck**

Figure 84

Figure 84   Peltigera malacea (×1)

Thallus greenish brown, turning deep green when wet, the lobes 20-30 mm wide, semierect, separating and falling apart when collected; lower surface whitish buff at the margin, darkening at the center, the veins barely distinguishable and rhizines mostly lacking; apothecia rare, erect. Common in open grassy areas in conifer forests. A related arctic species, *P. scabrosa* Th. Fr., has more distinct veins and does not turn green when wet.

PORED LICHENS

54a (45) Thallus with white pores (cyphellae or pseudocyphellae) on the lower or upper surfaces (use hand lens and compare with Figs. 6, 85, and 86) (if chestnut brown and with tiny pores, go on to 54b). 55

Figure 85

Figure 85   White pores (pseudocyphellae) of *Parmelia rudecta* (left) and *P. stictica* (×10)

54b   Thallus without white pores. 66

55a (54) Pores recessed in the lower surface, large enough to be seen without a lens (Fig. 86); no pores in upper surface. .. 56

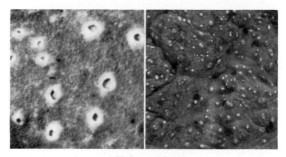

Figure 86

Figure 86   Cyphellae of *Sticta weigelii* (left) and pseudocyphellae of *Pseudocyphellaria anthraspis* (×10)

55b   Pores smaller, best seen with a hand lens, not recessed, either on the upper or lower surfaces (Fig. 86). 58

56a (55) Upper surface isidiate. Fig. 87. ....
.... *Sticta fuliginosa* (Dicks.) Ach.

Figure 87

Figure 87   Sticta fuliginosa

Thallus greenish to dark brown, thin, loosely attached, 5-10 cm broad; isidia usually clumped; lower surface tan, long-tomentose, with inconspicuous cyphellae; apothecia lacking. Rare on tree bases and mosses at higher elevations. The commoner S. *weigelii* is more leathery and has marginal soredia.

56b Margins and in part upper surface sorediate or isidiate. **57**

57a (56) Soredia granular to isidiate, marginal. Fig. 88. ....
.... *Sticta weigelii* (Ach.) Vain.

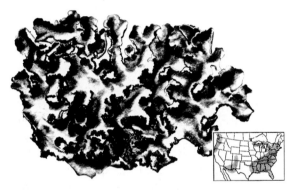

Figure 88

Figure 88   Sticta weigelii (×1)

Thallus brown to drab, loosely attached, 4-15 cm broad; lower surface light brown, tomentum dense; apothecia rare. Base of trees and on rocks in open woods, most common in the South. The cyphellae are conspicuous and make identification easy. Some specimens in the Western States may have coarse, mostly laminal isidia; these appear to be S. *sylvatica* (Huds.) Ach., a European species.

57b Soredia powdery, laminal and marginal. Fig. 89. .... *Sticta limbata* (Sm.) Ach.

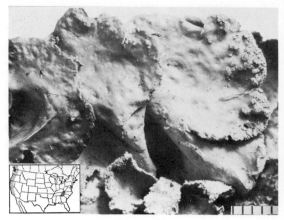

Figure 89

Figure 89   Sticta limbata

Thallus light brown, loosely attached, 4-10 cm broad; upper surface plane to somewhat ridged; lower surface buff, short-tomentose, cyphellae inconspicuous; apothecia lacking. Fairly common on bark or over mosses on trees.

58a (55) Pores on the lower surface only. ... **59**

58b Pores on the upper surface only. **61**

59a (58) Pores and soredia bright yellow. Fig. 90. ....
.. *Pseudocyphellaria aurata* (Ach.) Vain.

Figure 90

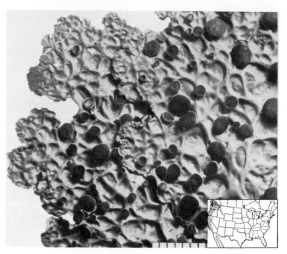

Figure 91

Figure 90   Pseudocyphellaria aurata

Thallus buffy olive, turning green when wet, loosely adnate, 4-10 cm broad; soredia marginal; lower surface pale tan, short tomentose; apothecia very rare. Conspicuous but not often collected, on trees in open deciduous forests. Related *P. crocata* (L.) Vain. has numerous small laminal soralia and occurs at higher elevations in the Appalachians, northern Great Lakes region, and the Pacific Northwest. Older books will list the *Pseudocyphellaria* species under *Sticta* but the pores are different.

59b   Pores and soredia (if present) white. ......
.... 60

60a (59) Thallus lacking soredia; apothecia usually well developed. Fig. 91. ....
.... *Pseudocyphellaria anthraspis* (Ach.) Gall. & James

Figure 91   Pseudocyphellaria anthraspis

Thallus brown to light brown, leathery, loosely attached, 6-20 cm broad; upper surface strongly ridged (without lens), shiny, without soredia; lower surface tan, short tomentose with numerous pores; apothecia common. On bark or rocks in open woods, often on conifer branches. Superficially it resembles *Lobaria pulmonaria* or *L. linita,* but these species have a mottled lower surface without pores. *Nephroma resupinatum* may key out here. It has large white papillae on the lower surface which resemble pores and large apothecia produced on the underside of the lobe tips (see page 131).

60b   Soredia present on the ridges; apothecia lacking. Fig. 92. ....
.... *Pseudocyphellaria anomala* Magn.

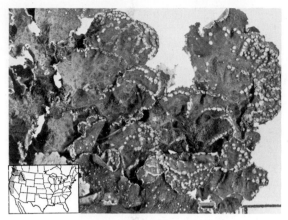

Figure 92

Figure 92   Pseudocyphellaria anomala

Thallus grayish brown, loosely attached, 6-10 cm broad; upper surface strongly ridged with soredia developing along the ridges; lower surface tan with short tomentum and numerous pores; apothecia very rare. On trunks of oak and on fallen logs in open forests. One other species of this genus, *P. rainierensis* Imsh., is known rarely from the Cascades of Washington and Oregon, usually growing on upper trunks of trees in the Douglas Fir forests. It is light mineral gray and has coarsely lobate-dentate margins.

61a (58) Soredia present on lobe margins and/or surface. **62**

61b   Soredia absent (isidia may be present). **64**

62a (61) Lower surface tan or pale brown. Fig. 93. .... *Parmelia subrudecta* **Nyl.**

Figure 93

Figure 93   Parmelia subrudecta (×1.5)

Thallus greenish mineral gray, adnate to loosely attached, 5-10 cm broad; soralia variable, laminal and/or marginal; lower surface pale brown, rhizinate nearly to the margin; apothecia very rare. Cortex K+ yellow (atranorin); medulla C+, KC+ red (lecanoric acid). Common on conifers and deciduous trees in open woods or along roadsides. This was formerly called *P. borreri* (Sm.) Turn., actually a rare species known from Ohio and West Virginia with a blackening lower surface and C+ rose test (gyrophoric acid). Another lecanoric acid-containing species in this group, *P. perreticulata* (Räs.) Hale, has narrow lobes (1-2 mm wide), a foveolate or ridged upper surface, and laminal soralia. It occurs in Texas and adjacent states.

62b   Lower surface black with a dark brown marginal zone. **63**

63a (62) Pores very small, not easily seen without a lens; medulla C+ red. Frontispiece No. 9 and Fig. 94. .... *Cetrelia olivetorum* (Nyl.) **Culb. & Culb.**

Figure 94

Figure 94   Cetrelia olivetorum (×2)

Thallus greenish mineral gray, loosely attached, 6-20 cm broad; pores 0.1-0.3 mm wide; lower surface with a broad bare zone along the margins; apothecia very rare. Cortex K+ yellow (atranorin); medulla C+, KC+ red (olivetoric acid). Common on trunks and large rock outcrops. Some specimens will react C−. For example, *C. monachorum* (Zahlbr.) Culb. & Culb., which contains predominantly imbricaric acid, is rather common in the Appalachians. Much less common is *C. cetrarioides* (Duby) Culb. & Culb., which contains more perlatolic acid and occurs in the Pacific Northwest as well. They can only be identified with thin-layer chromatography. *Cetrelia chicitae* (below) has the same range but differs in chemistry (C−) and in coarser soredia. Broad-lobed specimens of *Parmelia borreri* (see under *P. subrudecta*) and the brownish *P. stictica* (see page 120) may key out here. They have a much narrower bare zone below and different chemistry (gyrophoric acid).

63b   **Pores large, up to 1 mm in diameter, visible without a lens; medulla C−. Fig. 95.** ....
.. *Cetrelia chicitae* (Culb.) Culb. & Culb.

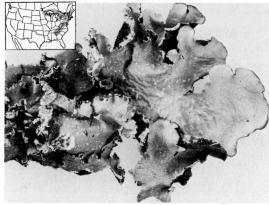

Figure 95

Figure 95   Cetrelia chicitae (×2)

Thallus greenish mineral gray, loosely attached, 6-20 cm broad; lobes broad and rotund, to 10 mm wide; soredia very coarse and granular; lower surface black at the center, brown or mottled white-brown at the margins; apothecia lacking. Cortex K+ yellow (atranorin); medulla K−, C−, KC+ red (alectoronic acid), P− (alectoronic acid). Common on rocks, more rarely trees, in open forests. Closely related *C. olivetorum* (above) has small pores and a C+ red reaction. A rare species that may key out here, *Parmelia reddenda* Stirt., will be found in the Great Smoky Mountains. It is smaller, has very coarse subisidiate soredia, and contains fatty acids (K−, C−, KC−, P−).

64a (61)   **Lower surface black. Fig. 96.** ....
.... *Parmelia appalachensis* Culb.

Figure 96

Figure 96   Parmelia appalachensis

Thallus whitish or pale greenish gray, adnate, 6-15 cm broad, rather brittle and fragile; upper surface wrinkled, becoming densely lobulate; lower surface sparsely rhizinate; apothecia occasional. Cortex K+ yellow (atranorin); medulla K−, C−, P− (fatty acids). Common on oak trees in open woods in the Appalachian Mountains. The black lower surface separates it from *Parmelia rudecta* and *P. bolliana*. In Texas there is another lobulate species, *P. subpraesignis* Nyl., which contains gyrophoric acid (C+ rose).

64b   Lower surface uniformly tan to light brown. **65**

65a (64) Upper surface isidate. Fig. 97. ....
.... *Parmelia rudecta* Ach.

Figure 97

Figure 97   Parmelia rudecta (×1.5)

Thallus greenish to bluish gray, adnate, rather brittle, 4-15 cm broad; upper surface becoming densely isidiate, the isidia variable, cylindrical to coarsely lobulate; lower surface densely rhizinate to the margin, tan; apothecia not common. Cortex K+ yellowish (atranorin); medulla K−, C+ red, P− (lecanoric acid). Very common on trees and rocks in open woods. This is one of the most frequently collected foliose lichens in eastern North America, often occurring with *Parmelina aurulenta* and *Pseudoparmelia caperata*.

65b   Upper surface without isidia. Fig. 98. ....
.... *Parmelia bolliana* Müll. Arg.

Figure 98

Figure 98 Parmelia bolliana

Thallus greenish mineral gray, adnate, 6-12 cm broad; upper surface becoming quite wrinkled, often lobulate with age; white pores inconspicuous; lower surface tan with moderate rhizines; black pores (pycnidia) (see Fig. 15B) and apothecia common. Cortex K+ yellow (atranorin); medulla K−, C−, P− (fatty acids). Common on tree trunks of exposed deciduous trees, especially in the prairie-forest border, or on sheltered acidic rocks. Some collections will react C+ red (lecanoric acid) and can be identified as *P. hypoleucites* Nyl. The two species occur together over most of the range.

66a (54) Thallus (when dry) pale mineral to greenish or whitish gray to white (see Frontispiece Nos. 8, 9, 10); surface of upper cortex K+ yellow (atranorin present; use hand lens for test) (exceptions: *Coccocarpia, Hypotrachyna formosana, Physciopsis syncolla, Placopsis gelida,* and *Pyxine caesiopruinosa*). 67

66b Thallus brown, ranging from tan to chestnut brown or darker or dark miner-

al gray (see Frontispiece Nos. 11 and 12) (a few species of *Physconia* may be white pruinose); cortex always K− or turning green as algae soak up water (atranorin lacking). 217

67a (66) Lobes generally broad, 4-20 mm wide (use ruler), apically rotund and with a more or less distinct bare zone below at the margins; thallus usually loosely attached to suberect (Figs. 3B, 26, and 99A). 68

Figure 99

Figure 99 Comparison of broad rotund lobes (left) and narrow linear lobes (about ×1)

67b Lobes generally narrow and linear, 0.5-6 mm wide, apically obtuse and without a distinct bare zone below at the margins (or lacking rhizines completely); thallus usually adnate to appressed (Figs. 3A, 99B) except for *Platismatia* species). ....
.... 97

BROAD-LOBED MINERAL GRAY LICHENS

68a (67) Soredia present on margins or surface of lobes. 69

68b Soredia lacking, the lobes isidiate or smooth. 83

69a (68) Black cilia present on lobe margins (use hand lens or hold specimen up to bright light, see Fig. 5A). 70

69b Cilia lacking on lobe margins. 77

70a (69) Lower surface with a broad, bare white zone 1 cm wide at the margin and a black center. Fig. 100. ....
.... *Parmotrema hypotropum* (Nyl.) Hale

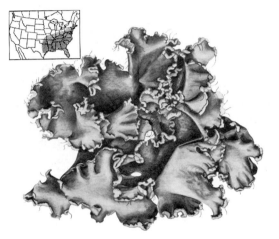

Figure 100

Figure 100 Parmotrema hypotropum (×1)

Thallus whitish to pale greenish mineral gray, loosely attached to suberect, 6-10 cm broad; apothecia lacking. Cortex K+ yellow (atranorin); medulla K+ yellow turning red, P+ orange (norstictic acid). Common on lower trunks and branches of hardwoods in open forests. A chemical variant with stictic and norstictic acids, *P. hypoleucinum* (B. Stein.) Hale, predominates on conifers in the coastal plain from Massachusetts to Mississippi and in southern California. Another rare variant, *P. louisianae* (Hale) Hale, reacts K−, P− (alectoronic acid) and occurs in Louisiana and Virginia but is probably more widespread. The white rim below in all the species separates them from *P. rampoddense* and *P. stuppeum.*

70b Lower surface dark brown (rarely mottled brown and white) in the marginal zone. 71

71a (70) Upper cortex finely reticulately cracked to the lobe margin (use hand lens and compare with Fig. 6A). Fig. 101. ....
Parmotrema reticulatum (Tayl.) Choisy

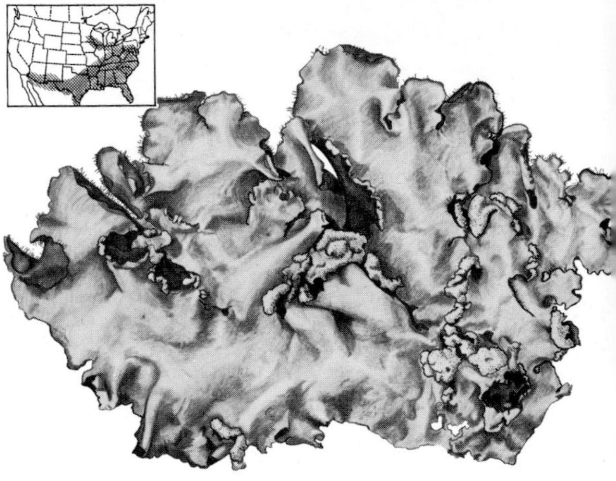

Figure 101

Figure 101 Parmotrema reticulatum (×3)

Thallus light mineral gray, adnate to loosely attached, 6-10 cm broad; soralia variable, powdery to pustulate and coarse, laminal to submarginal; lower surface sparsely rhizinate to the margin; apothecia rare. Cortex K+ yellow (atranorin); medulla K+ yellow→red, P+ orange (salazinic acid). Common in open woods and along roadsides. The reticulate cracking will separate it from *P. stuppeum* and *P. margaritatum,* both of which may appear to intergrade to some extent. Two rarely collected species in the southern Appalachians are very similar in external appearance but have different chemistry. *Parmotrema diffractaicum* (Essl.) Hale contains lichexanthone and diffractaic acid (K−, C−, P−) and *P. simulans* (Hale)

Hale, also negative with reagents, contains fatty acids.

71b Upper cortex continuous (irregularly cracked only on older lobes). 72

72a (71) Soredia laminal, rather diffuse; cilia very short, mostly in lobe axils. ....
.... (p. 80) *Parmelina aurulenta*

72b Soredia along the lobe margins; cilia longer, produced in the axils and on lobe tips. 73

73a (72) Collected in the Coastal Plain and Piedmont from Virginia to Texas. 74

73b Collected in the Appalachian-Great Lakes region and from Washington to California. 75

74a (73) Soredia produced in marginal soralia on main lobes; cilia very long, to 5 mm. Fig. 102. ....
.. *Parmotrema rampoddense* (Nyl.) Hale

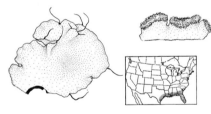

Figure 102

Figure 102   Parmotrema rampoddense (×2)

Thallus whitish mineral gray, loosely attached, 10-30 cm broad; lower surface sparsely rhizinate, naked and brown in a broad marginal zone; apothecia lacking. Cortex K+ yellow (atranorin); medulla KC+ red (alectoronic

acid). Common on palm and oak trees in open woods. There is some intergradation with specimens of *P. mellissii* which have exceptionally sorediate isidia.

74b Soredia produced mainly on small secondary lobes toward the thallus center; cilia shorter, produced irregularly in lobe axils. Fig. 103. ....
...... *Parmotrema dilatatum* (Vain.) Hale

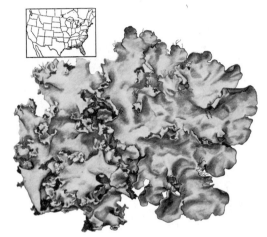

Figure 103

Figure 103   Parmotrema dilatatum (×1)

Thallus light mineral gray, adnate to loosely attached, 6-20 cm broad; margins becoming dissected and laciniate, rarely with sparse short cilia; lower surface with a broad bare zone along the margins; apothecia very rare. Cortex K+ yellow (atranorin); medulla K−, C−, P+ red (echinocarpic and protocetraric acids). Very common on trees in open forests, especially in Florida. If cilia are overlooked this lichen would be identified as *P. robustum*, which is virtually identical except for lacking echinocarpic acid. More often it will be confused with *P. cristiferum*, which is K+ red and has mostly marginal, linear soralia.

75a (73) Medulla K— (see Fig. 18 for section-ing technique). Fig. 104. ....
............*Parmotrema arnoldii* (DR.) Hale

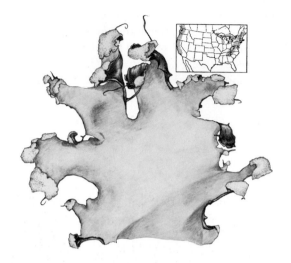

Figure 104

Figure 104   Parmotrema arnoldii (×10)

Thallus light mineral gray, loosely adnate, 6-10 cm broad; soralia mostly submarginal, often produced on short lobules or the dissected lobe margins; lower surface black with a narrow bare zone at the margins; apothecia lacking. Cortex K+ yellow (atranorin); medulla K—, KC+ red, C—, P— (alectoronic acid). Common on trees in open forests in the West but rather rare in the East. *Parmotrema margaritatum* (Hue) Hale has identical morphology but reacts K+ red (salazinic acid). Another species with a K— reaction, *P. hababianum* (Gyel.) Hale, has soredia in linear soralia; it has been collected so far only in Arizona.

75b Medulla K+ yellow or yellow turning red. 76

Figure 105

Figure 105   Position of soralia in *Parmotrema stuppeum* (left) and *P. perlatum* (enlarged)

76a (75) Lobes broad, to 10 mm wide, often suberect, with narrow marginal soralia (Fig. 105A). Fig. 106. ....
...... *Parmotrema stuppeum* (Tayl.) Hale

Figure 106

Figure 106   Parmotrema stuppeum

Thallus mineral gray, loosely adnate, 8-15 cm broad; lower surface with a broad naked marginal zone; apothecia very rare. Cortex K+ yellow (atranorin); medulla K+ yellow→red, P+ orange (salazinic acid). Common on deciduous trees in open woods or swamps, at higher elevations in the Appalachians. *Par-*

motrema margaritatum, a rare lichen from Iowa and Wisconsin to the Appalachians, has irregular round soralia on short marginal lobes (as in Fig. 104) and the same chemistry.

76b Lobes narrower and crowded, 4-8 mm wide with soredia produced on revolute lobe tips (Fig. 105B). Fig. 107. ....
.. *Parmotrema perlatum* (Huds.) Choisy

Figure 107

Figure 107   Parmotrema perlatum (×1.5)

Thallus mineral gray, loosely adnate, 5-10 cm broad; margins short ciliate; lower surface with a narrow bare zone along the margins; apothecia lacking. Cortex K+ yellow (atranorin); medulla K+ yellow, P+ orange (stictic acid). Quite common on oaks in California and Oregon, rarer in the eastern states. This species is most often confused with *P. arnoldii,* which reacts K− in the medulla and has mostly submarginal, smaller soralia.

77a (69) Collected in northern and western United States and Canada. 78

77b Collected in southeastern United States from North Carolina to Texas. 80

78a (77) Thallus mineral gray, the lobes suberect; soredia becoming granular or subisidiate on dissected lobe margins. Fig. 108. ....
.... *Platismatia glauca* (L.) Culb. & Culb.

Figure 108

Figure 108   Platismatia glauca

Thallus greenish mineral gray, loosely attached and suberect, 6-10 cm broad; upper surface plane to ridged with soredia scattered along the margin and in part isidiate; lower surface brown to mottled white and brown, nearly bare, shiny; apothecia lacking. Cortex K+ yellow (atranorin); medulla K−, C−, P− (fatty acids). On conifers in open forest, extremely common in the western states but rare in the East. The production of isidiate soredia is extremely variable.

78b Thallus yellowish green (usnic acid present), adnate; soredia powdery. .. 79

79a (78) Pores present in the upper cortex (use lens); medulla C+ red. .... **(p. 34)** *Parmelia flaventior*

79b Pores lacking; medulla C–. *Pseudoparmelia caperata* **(see page 34).**

80a (77) Medulla K+ red (see Fig. 18 for sectioning technique). Fig. 109. .... *Parmotrema cristiferum* **(Tayl.) Hale**

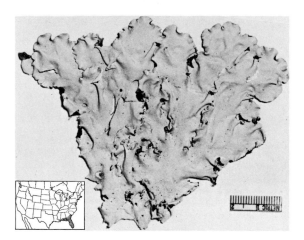

Figure 109

Figure 109 Parmotrema cristiferum

Thallus greenish to whitish mineral gray, loosely adnate, 10-20 cm broad; soredia produced in marginal soralia; lower surface with a broad brown bare zone; apothecia very rare. Cortex K+ yellow (atranorin); medulla K+ red, C–, P+ orange (salazinic acid). Common on oak and palm trees in open woods and along roads. The thallus may have a reddish tinge because of decomposition of the salazinic acid. In contrast *P. dilatatum* has a more whitish gray color, a K– medulla, and more irregular soralia produced on secondary lobes. *Parmotrema praesorediosum* (below) is much smaller and also reacts K–. One other confusable species, *P. dominicanum* (Vain.) Hale,

has been rarely collected in Florida; it has distinctly yellowish soredia on lobe margins and reacts K– in the medulla (protocetraric acid).

80b Medulla K–. **81**

81a (80) Lobes rather narrow, 3-5 mm wide, crowded; soralia typically produced in a crescent shape at lobe tips. Fig. 110. .... *Parmotrema praesorediosum* **(Nyl.) Hale**

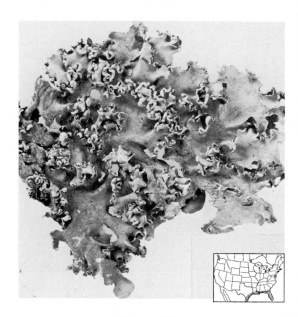

Figure 110

Figure 110 Parmotrema praesorediosum (×2)

Thallus greenish mineral gray, rather closely adnate, 5-10 cm broad; lower surface with a narrow bare zone along the margin; apothecia lacking. Cortex K+ yellow (atranorin); medulla K–, C–, P– (fatty acids). Common on oak trees in open woods and on citrus trees in Florida.

81b Lobes broader, to 10 mm wide; soralia produced mostly on secondary lobes toward the thallus center or on the margins. 82

82a (81) Medulla C+ red; lower surface often white along the margin. Fig. 111. .... *Parmotrema austrosinense* (Zahlbr.) Hale

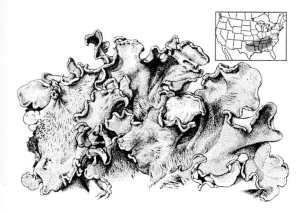

Figure 111

Figure 111   Parmotrema austrosinense (×1.5)

Thallus whitish mineral gray, loosely attached, 5-8 cm broad; lobes to 10 mm or more wide, usually suberect with a white margin below; apothecia lacking. Cortex K+ yellow (atranorin); medulla K−, C+ red, P− (lecanoric acid). Conspicuous, but rarely collected, on trees in open pastures and along roads. This lichen is similar to *P. hypotropum* but lacks cilia and has a different chemistry.

82b Medulla C−; lower surface brown to mottled brown and white along the margin. .... *Parmotrema robustum* (Degel.) Hale

Thallus morphology, habitats, and range as in *P. dilatatum* (above) but containing only protocetraric acid in the medulla. This species, which must be identified with chromatogra-phy, is collected much less frequently than *P. dilatatum*. Actually, it may also have cilia produced irregularly in the lobe axils.

83a (68) Isidia present (use hand lens and compare with Fig. 52). 84

83b Isidia absent, the lobes smooth to wrinkled. 91

84a (83) Medulla sulfur yellow to orange (expose with razor blade). 85

84b Medulla white. 86

85a (84) Medulla sulfur yellow; short cilia on lobe margins. .... (p. 37) *Parmotrema sulphuratum*

85b Medulla pale yellow orange; cilia completely lacking. Fig. 112. .... *Parmotrema endosulphureum* (Hillm.) Hale

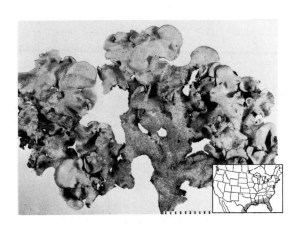

Figure 112

Figure 112   Parmotrema endosulphureum

Thallus greenish mineral gray, adnate to loosely attached on bark, to 15 cm broad, the lobes wide and rotund; medulla pale yellow orange; lower surface black with a wide bare brown zone at the margins; apothecia very rare. Cortex K+ yellow (atranorin); medulla K+ yellowish, C+ rose, P− (gyrophoric acid and pigments). Common on trees in open woods. This lichen has been confused with *P. sulphuratum* which has a deeper lemon yellow medulla and cilia. Externally it is very close to *P. tinctorum.*

86a (84) Lobe margins smooth, without cilia; medulla C+ red. Frontispiece No. 8 and Fig. 113. ....
.... ***Parmotrema tinctorum* (Nyl.) Hale**

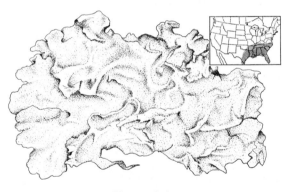

Figure 113

Figure 113   Parmotrema tinctorum (×2)

Thallus whitish mineral gray, adnate to loosely attached, 8-20 cm broad; isidia sparse to moderate, sometimes rather granular; lower surface with a broad bare brown zone; apothecia rare. Cortex K+ yellow (atranorin); medulla C+, KC+ (lecanoric acid). Very common on trees and rocks in open woods and along roadsides. This is one of the most frequently collected lichens in southern United

States. Larger forms of *Pseudoparmelia amazonica* which may key out here can be recognized by the C−, P+ red reaction. Another lichen, not included in this book, will key here: *Platismatia norvegica* (Lynge) Culb. & Culb., similar to *P. lacunosa* (see page 69) but isidiate. It has been collected rarely in Newfoundland and in the Cascades from Oregon into British Columbia.

86b   Lobe margins ciliate (use hand lens or hold up to bright light); medulla C−. ..
.... **87**

87a (86) Lower surface of thallus more or less uniformly brown; short rhizines present to lobe edge below. Fig. 114. ....
.... ***Parmotrema subtinctorium* (Zahlbr.) Hale**

Figure 114

Figure 114   Parmotrema subtinctorium

Thallus greenish mineral gray (sometimes turning buff in the herbarium), loosely adnate, 5-10 cm broad; upper surface densely isidiate, the cortex shiny and faintly white-spotted; lower surface rhizinate or papillate nearly to the margin; apothecia lacking. Cortex K+ yellow (atranorin); medulla K− or K+ yellow→ red, KC+ red (norlobaridone with or without salazinic acid). Fairly common on deciduous trees in open woods or along roadsides. The K− population without salazinic acid is a chemical strain called *P. haitiense* (Hale) Hale. The variant with salazinic acid alone is called *P. neotropicum* Kurok. These species could be mistaken for *P. crinitum* if the lower surface darkens but they can also be distinguished by the shiny, maculate upper cortex. *Parmotrema ultralucens* is a more robust plant with a dull upper cortex.

87b Lower surface black at the center and brown and more or less free of rhizines at the margin. **88**

88a (87) Medulla K− (see Fig. 18 for sectioning technique); isidia becoming sorediate. Fig. 115. ....
...... *Parmotrema mellissii* (Dodge) **Hale**

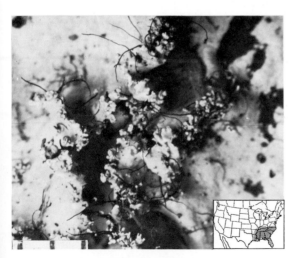

Figure 115

Figure 115 Parmotrema mellissii

Thallus whitish mineral gray, loosely adnate, 5-10 cm broad; isidia variable in density and development of soredia; lower surface sparsely rhizinate; apothecia lacking. Cortex K+ yellow (atranorin); medulla KC+ red (alectoronic acid). Widespread in open woods but not often collected. *Parmotrema crinitum* is roughly similar in external appearance but reacts K+ yellow in the medulla.

88b Medulla K+ yellow or yellow turning red; isidia not becoming sorediate. **89**

89a (88) Upper cortex finely reticulately cracked to the margin (use hand lens and compare with Fig. 6a). Fig. 116. ......
.... *Parmotrema subisidiosum* (Müll. Arg.) **Hale**

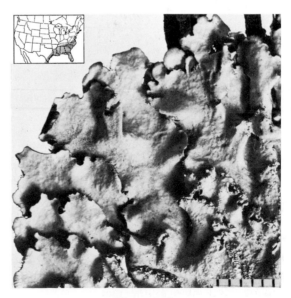

Figure 116

Figure 116 Parmotrema subisidiosum

Thallus light mineral gray, adnate, 4-10 cm broad; isidia rather irregular, becoming granular, often near the lobe margins; apothecia lacking. Cortex K+ yellow (atranorin); medulla K+ yellow→red, P+ orange (salazinic acid). Rather common on trees in open woods, especially in Florida. This species is closely related to *P. reticulatum,* which has the same reticulate cracking but is sorediate.

89b   Upper cortex continuous or irregularly cracked on older lobes. 90

90a (89) Medulla K+ persistent yellow; isidia dense, often branched and apically ciliate. Fig. 117. ....
...... Parmotrema crinitum (Ach.) Choisy

Figure 117

Figure 117   Parmotrema crinitum (×1.5)

Thallus whitish to greenish mineral gray, loosely adnate, 6-12 cm broad; isidia becoming branched and coralloid; lower surface with a narrow bare zone at the margin; apothecia rare. Cortex K+ yellow (atranorin); medulla K+ yellow, C−, P+ orange (stictic acid). Common on trunks of oaks, maples, and other hardwoods as well as cedar in open woods and on rocks. A smaller, narrow-lobed species with stictic acid and norlobaridone, *P. internexum* (Nyl.) Hale, is rarely collected in the southeastern states. It would be considered superficially as an abnormally small *P. crinitum.* *Parmotrema ultralucens* has broader, flatter lobes and salazinic acid (K+ red), and *P. subtinctorium* is brown below.

90b   Medulla K+ yellow turning red; isidia simple, moderately developed, without apical cilia. Fig. 118. ....
.... Parmotrema ultralucens (Krog) Hale

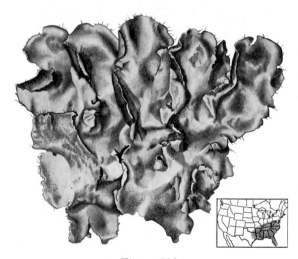

Figure 118

Figure 118   Parmotrema ultralucens (×2)

Thallus whitish mineral gray, loosely adnate, 5-15 cm broad; upper surface becoming whitish pruinose with age; lower surface sparsely rhizinate; apothecia lacking. Cortex K+ yellow (atranorin); medulla K+ red, C−, P+ orange (salazinic acid and lichexanthone). Common on deciduous trees and rocks in open woods. This species had previously been called *"Parmelia subcrinita."*

91a (83) Collected in the Pacific Northwest (California to Washington). Fig. 119. .... *Platismatia lacunosa* (Ach.) Culb. & Culb.

Figure 119

Figure 119 Platismatia lacunosa (×1)

Thallus mineral gray or whitish, loosely adnate, 4-10 cm broad; ridges sharp and distinct; lower surface white and black or brown mottled, bare; apothecia common. Cortex K+ yellow (atranorin); medulla P+ red (fumarprotocetraric acid). Fairly common on exposed conifers. The sharp ridges and geography separate this from *Platismatia tuckermanii*.

91b Collected in eastern and central North America. 92

92a (91) Cilia lacking (even in lobe axils) (use hand lens or hold up to bright light). 93

92b Cilia present on lobe margins and/or in axils (see Fig. 5A). 94

93a (92) Thallus suberect, bare and shiny below, often mottled. Fig. 120. .... *Platismatia tuckermanii* (Oakes) Culb. & Culb.

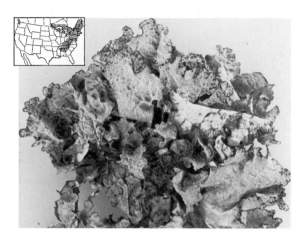

Figure 120

Figure 120 Platismatia tuckermanii (×2)

Thallus mineral gray to whitish, loosely adnate to suberect, 4-10 cm broad; lower surface white and black or brown mottled, sparsely rhizinate or bare; apothecia common. Cortex K+ yellow (atranorin); protolichesterinic acid also present in medulla. Common on branches and trunks of conifers in open woods. *Cetraria ciliaris* grows with this lichen very frequently and they should be carefully distinguished. Usually *C. ciliaris* has narrower brownish lobes and some marginal cilia with only weak ridging on the upper surface.

93b Thallus more adnate, the lobes not suberect, pale brown and short tomentose below (use lens and compare with Fig. 7). .... (p. 132) *Lobaria quercizans* and (p. 131) *L. ravenelii*

94a (92) Lower surface with a broad white rim; lobes often suberect. Fig. 121. ....
.. *Parmotrema perforatum* (Jacq.) Mass.

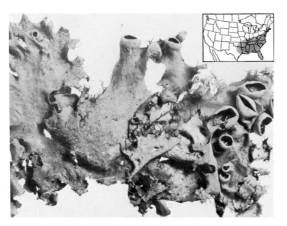

Figure 121

Figure 121   Parmotrema perforatum (×1.5)

Thallus whitish mineral gray, loosely attached to suberect, 6-12 cm broad; upper surface faintly white-spotted (use high power lens); lower surface bare to sparsely rhizinate; apothecia common, stalked and with a hole in the disc. Cortex K+ yellow (atranorin); medulla K+ yellow→red, P+ orange (norstictic and protolichesterinic acids). Common on branches at the tops of trees and on exposed trunks. A chemical variant with alectoronic acid (KC+ red), sometimes in combination with norstictic acid, is called *P. rigidum* (Lynge) Hale. It occurs in the Coastal Plain from North Carolina to Texas. Another variant, rarely collected but locally abundant in east Texas and Louisiana, *P. preperforatum* (Culb.) Hale, contains stictic, constictic, and norstictic acids.

94b   Lower surface brown at the margins; lobes adnate. 95

95a (94) Medulla K— (see Fig. 18 for sectioning technique). Fig. 122. .... *Parmotrema michauxianum* (Zahlbr.) Hale

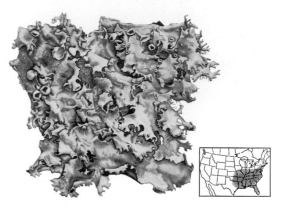

Figure 122

Figure 122   Parmotrema michauxianum (×1)

Thallus whitish mineral gray, adnate, 4-12 cm broad; pycnidia numerous on the upper surface; lobe axils with short cilia; lower surface moderately rhizinate; apothecia very common, somewhat stalked. Cortex K+ yellow (atranorin); medulla P+ red (protocetraric acid). Common on trunks and branches in open woods and along roadsides. The lobe margins may become quite dissected and laciniate.

95b   Medulla K+ yellow turning red. 96

96a (95) Upper cortex finely reticulately cracked to the margin (use hand lens and compare with Fig. 6A). Fig. 123. .... *Parmotrema cetratum* (Ach.) Hale

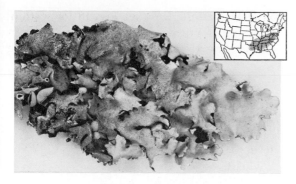

Figure 123

Figure 123   Parmotrema cetratum (×1.5)

Thallus light mineral gray (often turning reddish after pressing and drying), adnate, 6-12
cm broad; lobes often laciniate; cilia rather inconspicuous; lower surface densely rhizinate
but with a narrow marginal zone; apothecia
common, the disc with a central hole. Cortex
K+ yellow (atranorin); medulla K+ yellow→
red, P+ orange (salazinic acid). Widespread
on deciduous trees in open woods or pastures
but not commonly collected. This is the nonsorediate "parent" of *P. reticulatum* and *P. subisidiosum*. *Parmotrema perforatum* also has
perforate apothecia but is white below at the
margin and contains norstictic acid.

**96b   Upper cortex continuous, cracked only
on older lobes. Fig. 124.** ....
.... *Parmotrema eurysacum* (Hue) Hale

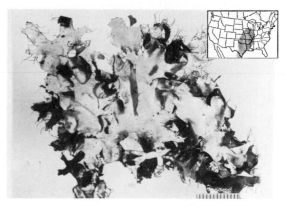

Figure 124

Figure 124   Parmotrema eurysacum

Thallus mineral gray, rather loosely adnate,
up to 10 cm broad, the lobes broad, adnate,
sometimes developing laciniae toward the center; lower surface brown to mottled at the margin; apothecia common, the disk perforate.
Cortex K+ yellow (atranorin); medulla K+
yellow turning red, C−, P+ orange (salazinic
acid). Widespread, but not commonly collected, on trees in open areas. Many specimens
of this rather undistinguished lichen had previously been identified as *P. perforatum* because of the perforate disc, but the lower surface color and chemistry are different.

NARROW-LOBED MINERAL GRAY LICHENS WITH A BLACK LOWER SURFACE

**97a (67)   Lower surface of thallus jet black,
at least at the center, and black to dark
brown (rarely mottled brown or white-
brown) at the margin (use lens only to
examine tiny specimens).** 98

**97b   Lower surface uniformly brown to tan
or white (orange pigmented only in**

Heterodermia obscurata and darkening at the center in some ecorticate *Heterodermia* species). 175

98a (97) Thallus lobes inflated and puffy, hollow (section with razor blade and compare with Fig. 125); lower surface completely lacking rhizines. 99

Figure 125

Figure 125   Section of a hollow lobe of *Hypogymnia*

98b   Thallus lobes flattened, the surface plane to convex (appearing inflated only in *Anzia colopodes* and *Hypogymnia oroarctica*); lower surface usually rhizinate, rarely tomentose or bare. 108

99a (98) Soredia present on surface or tips of lobes; apothecia rare. 100

99b   Soredia lacking; apothecia often present. 103

100a (99) Upper surface more or less regularly perforated with holes. Fig. 126. ....
.. *Menegazzia terebrata* (Hoffm.) Mass.

Figure 126

Figure 126   Menegazzia terebrata (×2)

Thallus light mineral gray, adnate, 6-15 cm broad; soredia powdery, often originating around the holes; lower surface deeply wrinkled, tearing away when thallus is removed from bark; apothecia very rare. Cortex K+ yellow (atranorin); medulla K+ yellow, P+ pale orange (stictic acid). Relatively rare on roadside trees and in swamps and bogs. This remarkable lichen was formerly known as *Parmelia pertusa.*

100b   Upper surface entire, rarely with a few irregular perforations. 101

101a (100) Soredia in labriform soralia on lower side of lobe tips which break open. Fig. 127. ....
.... *Hypogymnia physodes* (L.) Nyl.

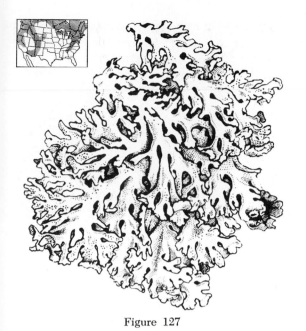

Figure 127

Figure 127 Hypogymnia physodes (×1.5)

Thallus light mineral gray, adnate to loosely attached, 6-12 cm broad; lobes appearing inflated; soralia more or less labriform; lower surface smooth to wrinkled and irregularly lacerated and perforated; apothecia very rare. Cortex K+ yellow (atranorin); medulla KC+, P+ red (physodic acid, physodalic acid, and protocetraric acid). Very common on conifers, hardwoods, and fenceposts throughout northern forests. Two rather similar species occur in the Appalachians: *H. appalachensis* Pike has similar soralia but is much smaller (2-4 cm broad) and darker, greener mineral gray (it is the sorediate morph of *H. krogii*); and *H. vittata* (Ach.) Gas. has sparse apical soredia, lobes with a conspicuous black margin, and small adventitious side branches. It reacts P−. *H. tubulosa* (below) has a different type of soralium and also reacts P−.

101b Soredia produced on surface of lobes or at the tips, not in labriform soralia; tips not breaking open. 102

102a (101) Soredia on short suberect lobes, appearing ring-shaped. Fig. 128. ....
.... *Hypogymnia tubulosa* (Schaer.) Hav.

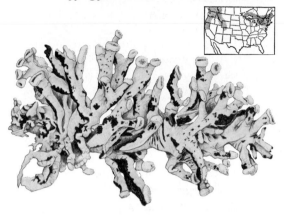

Figure 128

Figure 128 Hypogymnia tubulosa (×2)

Thallus mineral gray, loosely attached with semi-erect lobes, 4-6 cm wide; lobes rather short with little branching, 2-3 mm wide with rare perforations; medullary cavity completely black; soralia terminal, appearing ring-shaped; lower surface black or dark brown, rugose; apothecia lacking. Cortex K+ yellow (atranorin); medulla K−, C−, P− (physodic acid). Widespread on conifers in open woods. The terminal soralia, short lobes, and chemistry separate this species from the much more common *H. physodes*.

102b Soredia mostly subterminal or laminal in round or diffuse soralia. Fig. 129. ....
.... *Hypogymnia bitteri* (Lynge) Ahti

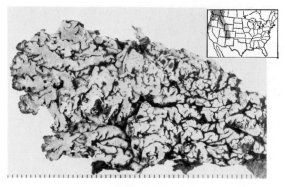

Figure 129

Figure 129   Hypogymnia bitteri

Thallus tannish to brownish mineral gray to brown, adnate on bark, 5-10 cm broad; lobes shiny with black-lined cracks in the upper surface; soralia mostly subterminal, powdery; medullary cavity blackening; lower surface rugose, black; apothecia lacking. Cortex K+ yellow (atranorin); medulla K−, C−, P− (physodic acid). Common on conifers at higher elevations in mountains. A related species, *H. austerodes* (Nyl.) Räs., has mostly diffuse laminal soralia and is more common in Canada. Another laminally sorediate species, *H. pseudophysodes* (Asah.) Rass., has no tinge of brown and also reacts P−. It has been collected a few times in Oregon north into British Columbia.

103a (99) Collected in the Appalachian Mountains. Fig. 130. ....
.... *Hypogymnia krogii* Ohlss.

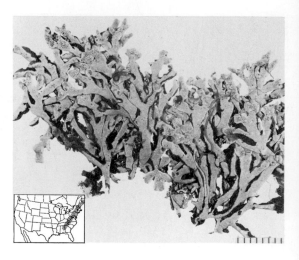

Figure 130

Figure 130   Hypogymnia krogii

Thallus light mineral to greenish gray, loosely attached, often forming cushion-like colonies, 5-8 cm broad; lobes narrow, rather short and adnate, the tips inflated, 1-2 mm wide; upper medullary cavity white, the lower black; lower surface rugose, dark brown to black; apothecia common. Cortex K+ yellow (atranorin); medulla K−, C−, P+ red. (physodalic and physodic acids). Rather common on twigs and branches of conifers at high elevations southward or in exposed sites in New England and north. This species was formerly called *Hypogymnia enteromorpha,* a western lichen with different morphology and chemistry.

103b Collected in western North America. ......
.... **104**

104a (103) Medullary cavity completely white (cut several lobes open with a razor blade). Fig. 131. ....
.... *Hypogymnia imshaugii* Krog

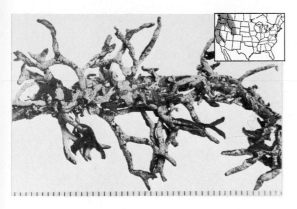

Figure 131

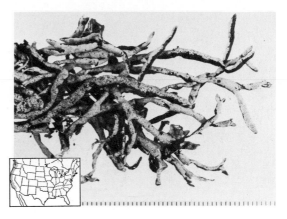

Figure 132

Figure 131   Hypogymnia imshaugii

Thallus mineral gray, extremely variable, loosely attached with free lobe ends to quite adnate, 5-9 cm broad; lobes rather short, 1.5-2 mm wide; upper surface heavily black spotted (pycnidia); lower surface very wrinkled, black to dark brown; apothecia abundant, short-stalked. Cortex K+ yellow (atranorin); medulla K−, C−, P+ red (physodic, physodalic, and often diffractaic acid) or P− (physodic only). Widespread and commonly collected on conifers. The P− population often occurs at high elevations. A rarer species, *H. metaphysodes* (Asah.) Rass., which is P−, has the upper roof of the medullary cavity white, the lower dark.

104b Medullary cavity dark brown to black above and below. **105**

105a (104) Thallus loosely attached at the base with rather long, free or trailing, narrow lobes. Fig. 132. ....
.... *Hypogymnia heterophylla* Pike

Figure 132   Hypogymnia heterophylla

Thallus whitish gray, 8-15 cm long; lobes variable, 1-4 mm wide, quite long, the apices perforate, margins black-rimmed; lower surface shiny, wrinkled, jet black. Cortex K+ yellow (atranorin); medulla K−, C−, P+ red (physodic and physodalic acids). Common on conifers along the Pacific coast. A comparable species, *H. duplicata* (Ach.) Rass., has no perforations and occurs primarily in coastal Canada but also in Oregon and Washington. A P− species, *H. inactiva* (Krog) Ohlss., is smaller but still trailing with uniformly thickened lobes. It occurs from California to Canada.

105b Thallus more or less adnate to bark, the lobes relatively short and wide. **106**

106a (105) Upper surface strongly rugose-pitted (without lens). Fig. 133. ....
.... *Hypogymnia rugosa* (Merr.) Pike

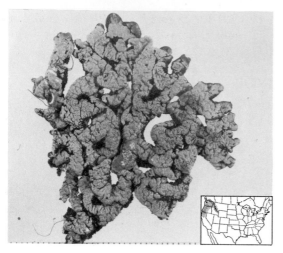

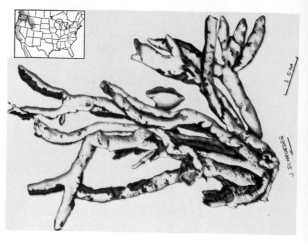

Figure 133

Figure 134

Figure 133   Hypogymnia rugosa (×1.5)

Thallus whitish mineral gray (turning buff in the herbarium), 5-25 cm broad; lobes broad and sparsely branched, 2-6 mm wide; medullary cavity entirely blackened; lower surface wrinkled, black, with irregular pores; apothecia common. Cortex K+ yellow (atranorin); medulla K−, C−, P− (hypoprotocetraric acid). Infrequent, on upper trunks of conifers. This unusual species is most common in the Cascades but rare eastward in the Rockies.

106b Upper surface smooth. 107

107a (106) Lobes strongly inflated, to 5 mm wide with a black rim when viewed from above. Fig. 134. ....
Hypogymnia enteromorpha (Ach.) Nyl.

Figure 134 Hypogymnia   enteromorpha (×1.5)

Thallus whitish mineral gray, loosely attached but not trailing, 6-12 cm broad; lobes to 5 mm wide; medullary cavity dark; lower surface dull, rugose, black, sporadically perforated; apothecia common, stalked. Cortex K+ yellow (atranorin); medulla K−, P− (physodic acid) or P+ red (physodalic and protocetraric acids). Very common on *Sequoia*, redwoods, and other conifers. This is the commonest *Hypogymnia* in the Cascades, becoming much rarer in the Rockies.

107b Lobes smaller, 2-3 mm wide, lacking a conspicuous black rim. Fig. 135. ....
.... *Hypogymnia occidentalis* Pike

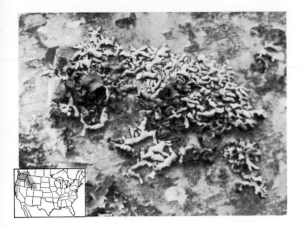

Figure 135

Figure 135 Hypogymnia occidentalis (×1)

Thallus whitish mineral gray, rather closely adnate, 5-10 cm broad; lobes rather short and irregularly inflated; medullary cavity entirely dark; lower surface deeply rugose, black, sparsely perforate. Cortex K+ yellow (atranorin); medulla K−, C−, P− (physodic acid). Common on main trunks of conifers in forests. This species, most commonly collected east of the Cascades, is related to *H. enteromorpha* and can be differentiated by the closer adnation and lack of a black rim.

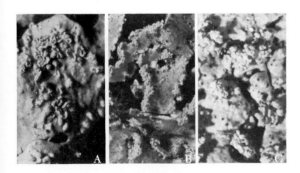

Figure 136

Figure 136 Examples of soredia in *Parmelina aurulenta* (A, pustulate), *Physcia crispa* (B, linear), and *Hypotrachyna formosana* (C, pustulate) (all about ×10)

108a (98) Lobes sorediate with marginal, laminal, or apical soralia or pustulate with erupting pustules resembling soredia (see Fig. 136); apothecia very rare. .. 109

108b Lobes isidiate or lobulate or smooth. 137

109a (108) Medulla white (use razor blade and see Fig. 18 for sectioning technique). 110

109b Medulla pale orange, pale yellow, or red. 133

110a (109) Upper surface weakly reticulately ridged and white-spotted (examine lobe tips carefully with hand lens and see Fig. 137). 111

Figure 137

Figure 137 White markings of *Parmelia sulcata* (left) and reticulation of *Pyxine eschweileri* (about ×10)

110b Upper surface plane to convex (ridged only in *Pseudoparmelia crozalsiana*), lacking white markings and spots. 113

111a (110) Collected in the Coastal Plain from South Carolina to Alabama and in Florida. Fig. 138. ....
.... *Pyxine eschweileri* (Tuck.) Vain.

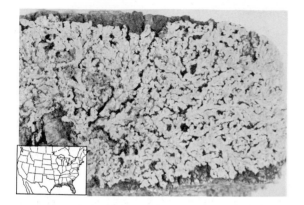

Figure 138

Figure 138   Pyxine eschweileri (×2)

Thallus whitish mineral gray, closely adnate on bark, 3-5 cm broad; lobes 0.5-1 mm wide, crowded, the upper surface cracked with reticulate white markings at the tips; soredia developing along margins and surface of lobes; lower surface black, rhizinate; apothecia rare. Cortex K+ yellow (atranorin); medulla K+ pale yellow, C−, P+ yellowish (unidentified substances). Common on oaks and magnolia. This is the commonest *Pyxine* with a white medulla. Imshaug's revision of *Pyxine* (1957) should be consulted for other species.

111b Collected north of the Coastal Plain. ....
.... **112**

112a (111) Soredia mostly laminal, often along ridges; rhizines densely branched. Fig. 139. .... *Parmelia sulcata* Tayl.

Figure 139

Figure 139   Parmelia sulcata

Thallus whitish mineral gray, adnate, 3-10 cm broad; lower surface black and densely rhizinate, the rhizines squarrosely branched (under lens); apothecia not common. Cortex K+ yellow (atranorin); medulla K+ yellow→red, P+ orange (salazinic acid). Very common on trees and rocks along roadsides or in woods. *Parmelia saxatilis* should be carefully distinguished by the coarse isidia. Both species often turn reddish after wetting and pressing for the herbarium.

112b Soredia mostly marginal; rhizines simple, unbranched. Fig. 140. ....
.... *Parmelia fraudans* Nyl.

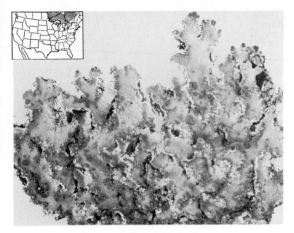

Figure 140

Figure 140 Parmelia fraudans (×2)

Thallus greenish mineral gray with a yellowish cast, adnate on rocks, 3-10 cm broad; lower surface black and densely rhizinate; apothecia lacking. Cortex K+ yellow (atranorin); medulla K+ yellow→red, P+ orange (salazinic acid). Common on exposed rocks near lakeshores but rarer inland. The weaker white markings and entirely marginal soralia separate this lichen from the much more common *P. sulcata*.

113a (110) Soralia mostly marginal (in part apical) and linear (Fig. 136A). 114

113b Soralia mostly laminal or apical, round or diffuse or pustular (Fig. 136B, C). 116

114a (113) Lobes rather broad, 3-6 mm wide, the margins subascending; soralia crescent-shaped. .... *Parmotrema praesorediosum* (see page 64)

114b Lobes narrow and linear, 1-2 mm wide, adnate. 115

115a (114) Upper surface scabrid and becoming white pruinose (use hand lens and compare with Fig. 6D); cortex K−. .... *Physconia detersa* (see page 119)

115b Upper surface smooth, only in part faintly pruinose; cortex K+ yellow. Fig. 141. .... *Physcia crispa* Nyl.

Figure 141

Figure 141 Physcia crispa (×4)

Thallus whitish gray, adnate, fragile, 4-6 cm broad; lobe margins white rimmed, becoming sorediate; lower surface ecorticate, white at the margin but turning blackish at the center, moderately rhizinate; apothecia rare. Cortex and medulla K+ yellow (atranorin and zeorin). Widespread at the base of trees in mature woods in the southern states. A confusable species, *Heterodermia albicans* (see page 105), has a shiny corticate lower surface and K+ red medullary test. *Anzia ornata* (Zahlbr.) Asah., though quite unrelated, will key out here. It has dense black tomentum below, as in *A. colpodes* (see Fig. 171), and reacts K− (divaricatic acid). It occurs very rarely in North Carolina and Alabama.

116a (113) Collected on trees. **117**

116b Collected on rocks. **129**

117a (116) Lobes 1-5 mm wide; thallus rather loosely adnate and usually easily removed from the substrate. **118**

117b Lobes narrow, 0.3-1.5 mm wide; thallus appressed, not easily removed without bark. **127**

118a (117) Rhizines simple, unbranched or sparsely furcate (examine carefully with a hand lens and compare with Fig. 7). ....
.... **119**

118b Rhizines becoming richly dichotomously branched. **125**

119a (118) Upper surface broadly wrinkled and reticulately ridged (use low power lens or no lens), sorediate along the ridges; medulla K+ yellow. Fig. 142. ....
.... *Pseudoparmelia crozalsiana* **(Lesd.) Hale**

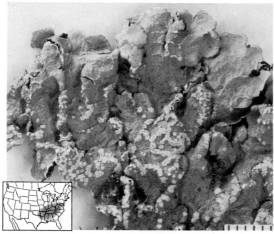

Figure 142

Figure 142   Pseudoparmelia crozalsiana

Thallus light mineral gray, adnate, 4-7 cm broad; lobes rather broad, to 5 mm wide, the upper surface foveolate and white reticulate, especially toward the tips; cilia lacking on lobe margins; lower surface black at the center, brown at the margin, densely rhizinate; apothecia lacking. Cortex K+ yellow (atranorin); medulla K+ yellow, C−, P+ orange (stictic acid). Widespread, but not commonly collected, on hardwood trees. The far more common *Parmelina aurulenta*, while superficially similar, has short cilia in the axils, diffuse soralia, and a pale yellow medulla.

119b Upper surface plane to irregularly wrinkled; soralia scattered or apical; medulla K− (K+ yellow only in pigmented *Parmelina aurulenta* and *Heterodermia casariettiana*). **120**

120a (119) Soralia broad and diffuse on the surface; pale pigment often visible in medulla, especially under soralia (use hand lens). Fig. 143. ....
.... *Parmelina aurulenta* **(Tuck.) Hale**

Figure 143

Figure 143 Parmelina aurulenta

Thallus greenish to bluish mineral gray, adnate, 4-10 cm broad; soredia sometimes arising from coarse pustules (see Fig. 136C) but becoming powdery; margins and/or lobe axils short cilia; lower surface densely rhizinate to the margin; apothecia rare. Cortex K+ yellow (atranorin); medulla K+, C+, P− (terpenes and if pigmented, entothein). Very common on trees and rocks in woodlots and along roads. This is one of the most commonly collected lichens in eastern North America, often occurring with *Parmelia rudecta* and *Pseudoparmelia caperata*. A superficially similar species, *Pseudoparmelia crozalsiana* (above), can be distinguished by the ridging, lack of cilia, and different chemistry.

120b Soralia more delimited and smaller, round or linear; medulla white. **121**

121a (120) Lobes 1-3 mm wide, narrow and linear, usually with some rhizines projecting out from below (used hand lens). **122**

121b Lobes 3-6 mm wide, irregularly broadened, no rhizines visible from above. **123**

122a (121) Lower cortex absent, the lower surface cottony white at the tips, purple black at the center. .... **(p. 104) *Heterodermia casarettiana***

122b Lower cortex present, uniformly blackened. .... **(p. 124) *Phaeophyscia hispidula***

123a (121) Soralia conspicuous, marginal, capitate and elevated. Fig. 144. .... ***Pseudoparmelia cryptochlorophaea* (Hale) Hale**

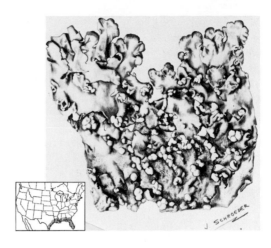

Figure 144

Figure 144 Pseudoparmelia cryptochlorophaea (×1)

Thallus greenish mineral gray, closely adnate, 5-10 cm broad; upper surface often white-reticulate at the lobe tips; lower surface black and rhizinate but turning brown and naked in a narrow marginal zone; apothecia very rare. Cortex K+ yellow (atranorin); medulla KC+ red (cryptochlorophaeic acid). On deciduous trees along roadsides and in moist woods from Texas to Florida in the Coastal Plain.

123b Soralia laminal or apical, not elevated. **124**

124a (123) Soralia apical or submarginal; some cilia present (at least in lobe axils); medulla K+ yellow or K−. .... **(p. 62) *Parmotrema arnoldii*, (p. 63) *P. perlatum*, or (p. 60) *P. reticulatum***

124b Soralia laminal, cilia completely lacking; medulla K—. Fig. 145. ....
.... *Pseudoparmelia texana* (Tuck.) Hale

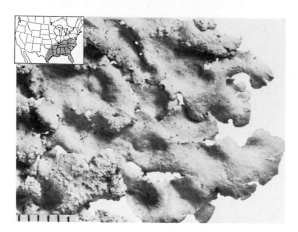

Figure 145

Figure 145   Pseudoparmelia texana

Thallus whitish mineral gray, adnate, 6-12 cm broad; upper surface becoming faintly white-reticulate at the tips; lower surface sparsely rhizinate with a narrow bare zone along the margins; apothecia very rare. Cortex K+ yellow (atranorin); medulla KC+ faint purple (divaricatic acid). Rather rare on conifers and hardwoods in dry open woods or on roadsides. A smaller saxicolous species of similar appearance, but with lobes only 1 mm wide, is *P. alabamensis* (Hale & McCull.) Hale; it reacts P+ red (protocetraric acid) and is known only in northern Alabama and Tennessee on rocks.

125a (118) Soredia laminal, arising from inflated, hollow pustules or only pustules present (use hand lens and compare with Fig. 136 B). Fig. 146. ....
Hypotrachyna formosana (Zahlbr.) Hale

Figure 146

Figure 146   Hypotrachyna formosana

Thallus whitish gray, closely adnate, 3-8 cm broad; upper surface becoming densely pustulate, the pustules usually erupting without forming soredia; apothecia lacking. Cortex K— (lichexanthone, fluorescing bright orange under an ultraviolet lamp); medulla K— or reddish, C—, P— (lividic acid group). On pines and deciduous trees in open forests, not commonly collected. Some specimens with identical morphology will lack fluorescence and contain atranorin in the cortex; these are *H. pustulifera* (Hale) Hale, which occurs throughout central and southern United States. A very rare pustular species occurring in Ohio and Pennsylvania, *H. showmanii* Hale, can be separated by chemistry (medulla KC+ red, unknown substances); it will not fluoresce either. Finally, *H. croceopustulata* (Kurok.) Hale, rather common in the southern Appalachians, is coarsely pustulate with a red pigment under the pustules; it reacts K—, P+ red (protocetraric acid).

125b Soredia powdery, mostly subapical, pustules indistinct or lacking. 126

126a (125) Rhizines richly dichotomously branched (use hand lens and see Fig. 484); soralia subterminal with powdery soredia and only a few erupting pustules present. Fig. 147. .... *Hypotrachyna rockii* (Zahlbr.) Hale

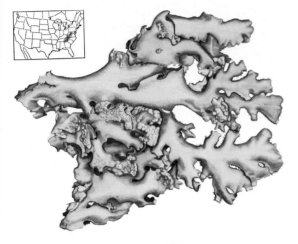

Figure 147

Figure 147   Hypotrachyna rockii (×2)

Thallus whitish mineral gray, adnate, 5-12 cm broad; lower surface black and densely rhizinate; apothecia lacking. Cortex K+ yellow (atranorin); medulla K−, C+ rose or C−, P− (evernic and lecanoric acids). Common on trunks of conifers in the Southern Appalachians. There are a number of "chemical species" that resemble *H. rockii* very closely and must be identified with chromatography. For example, virtually identical *H. gondylophora* (Hale) Hale is K−, P+ red (fumarprotocetraric acid), *H. laevigata* (Sm.) Hale K−, C+, KC+, orange, P− (barbatic acid group),

H. oostingii (Dey) Hale K−, C+ rose (gyrophoric acid and an unknown substance), *H. producta* Hale K−, C+ red, P− (anziaic acid), and *H. thysanota* (Kurok.) Hale K−, C+ rose, P+ yellow (gyrophoric, microphyllinic, and echinocarpic acids). These species are locally common but in general dominated by *H. rockii*.

126b Rhizines sparsely dichotomously branched; soralia with coarse, eroding soredia and open pustules, the sorediate lobes turning under. Fig. 148. .... *Hypotrachyna revoluta* (Flk.) Hale

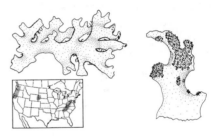

Figure 148

Figure 148   Hypotrachyna revoluta (×5)

Thallus mineral gray, adnate but with lobe tips ascending, 3-7 cm broad; lobes 1-2 mm wide; lower surface moderately rhizinate; apothecia lacking. Cortex K+ yellow (atranorin); medulla K−, C+ rose, P− (gyrophoric acid). Rather rare on trees in open areas or on sheltered rocks at higher elevations. In the *Rhododendron* glades of the Southern Appalachians one may collect *Everniastrum catawbiense* (Degel.) Hale, which resembles *H. revoluta* in general aspect but differs in having long marginal cilia and convoluted lobes.

127a (117) Thallus closely adnate, greenish to brownish mineral gray; cortex K–. .... **(p. 122)** *Phaeophyscia species*

127b Thallus adnate, whitish gray; cortex K+ yellow. **128**

128a (127) Rhizines lacking below (use hand lens). Fig. 149. .... *Dirinaria picta* **(Sw.) Clem. & Schear**

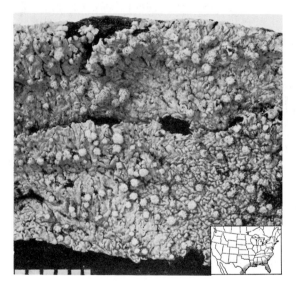

Figure 149

Figure 149 Dirinaria picta

Thallus whitish mineral gray, closely adnate, 3-5 cm broad with the colonies often fusing; soralia numerous; apothecia rare. Cortex K+ yellow (atranorin); medulla K–, C–, P– (divaricatic acid). Common on deciduous trees in open woods, citrus groves, and along roadsides. *Dirinaria aspera,* a closely related species, has pustular, almost sorediate isidia.

Physcia sorediosa, which may key here, has rhizines on the lower surface. A rare, unrelated western lichen may key here as well: *Cavernularia hultenii* Degel., which has conspicuous deep pits or pores in the lower surface. It is the sorediate morph of *C. lophyrea* (see page 98).

128b Rhizines present on the lower surface. Fig. 150. .... *Physcia sorediosa* **Vain.**

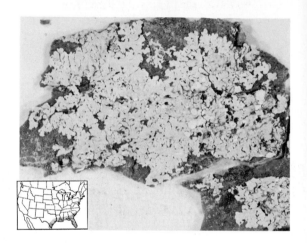

Figure 150

Figure 150 Physcia sorediosa (×2)

Thallus whitish gray, closely adnate on bark, 2-4 cm broad; lobes crowded, about 1 mm wide; soralia round, mostly laminal; lower surface black, moderately rhizinate; apothecia lacking. Cortex and medulla K+ yellow (atranorin and zeorin). Widespread on oak and other hardwoods in open areas. This species had previously been confused with *P. americana,* which has a uniformly tan lower surface and a different terpene.

129a (116) Lobes irregular, 1-5 mm wide; soredia becoming diffuse; medulla sometimes pale yellow orange under the soralia. ....
.... **(p. 80)** *Parmelina aurulenta*

129b Lobes linear and narrow, 0.5-1-5 mm wide. **130**

130a (129) Rhizines present on the lower surface or margins (use hand lens). **131**

130b Rhizines lacking. **132**

131a (130) Lower surface uniformly black, corticate, with branched rhizines. ....
.... **(p. 83)** *Hypotrachyna revoluta*

131b Lower surface lacking a cortex, black at the center and cottony white at the margins with long marginal rhizines. ....
.... **(p. 104)** *Heterodermia casarettiana*

132a (130) Collected in the Pacific Northwest. Fig. 151. ....
.... *Placopsis gelida* **(L.) Linds.**

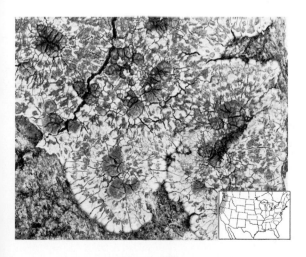

Figure 151

Figure 151   Placopsis gelida

Thallus greenish mineral gray, appressed, chinky-crustose in the center and lobate at the margins, 2-4 cm broad; surface with large tan colored warts (cephalodia), 1-3 mm wide; soralia becoming worn away, leaving pits; apothecia common. Medulla C+, KC+ red (gyrophoric acid). Common on rocks in open areas. This is a pioneer lichen often seen on moist rocks near rivers or moraines.

132b Collected in eastern United States. Fig. 152. .... *Dirinaria frostii* **(Tuck.) Awas.**

Figure 152

Figure 152   Dirinaria frostii

Thallus whitish mineral gray, tightly appressed on rock, 2-3 cm broad; soralia orbicular, laminal; apothecia lacking. Cortex K+ yellow (atranorin); medulla K−, C−, P− (divaricatic acid). On ledges in open woods, often on overhanging surfaces. This rarely collected lichen prefers flat surfaces and is often impossible to remove. *Pseudoparmelia alabamensis* (Hale & McCull.) Hale may key out here. It has rhizines and a P+ medulla (protocetraric acid).

133a (109) Lobes 2-6 mm wide, adnate to loosely adnate, the soralia (or pustules) diffuse. .... (p. 82) *Hypotrachyna croceopustulata* and *Parmelina aurulenta*

133b Lobes 1-2 mm wide, closely adnate to appressed, the soralia small, linear or round. 134

134a (133) Medulla deep red, K+ purple. .... (p. 118) *Phaeophyscia rubropulchra*

134b Medulla pale orange or yellowish, K+ yellow. 135

135a (134) Soredia mostly laminal in round soralia. Fig. 153. .... *Pyxine caesiopruinosa* (Nyl.) Imsh.

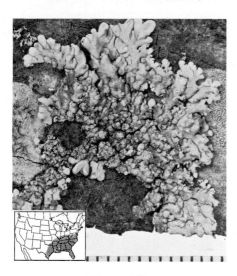

Figure 153

Figure 153   Pyxine caesiopruinosa

Thallus whitish mineral gray, closely adnate, 2-3 cm broad; lobe tips often with a distinct patch of white pruina (under lens); lower surface black and sparsely rhizinate; apothecia

very rare. Cortex brilliant orange in UV light (lichexanthone); medulla K+, C+, KC+ yellow, P− (sekikaic acid and pigments). Common on deciduous trees in pastures and along roads. Formerly confused with *P. sorediata*, it has mostly round laminal soralia and a different chemistry.

135b Soralia mostly marginal and linear. .. 136

136a (135) Medulla salmon orange; lobe margins white and splitting. Fig. 154. .... *Pyxine sorediata* (Ach.) Mont.

Figure 154

Figure 154   Pyxine sorediata

Thallus whitish to greenish mineral gray, closely adnate, 3-8 cm broad; upper surface becoming scabrid or lightly pruinose; lower surface black, densely rhizinate; apothecia very rare. Cortex K+ yellow (atranorin); medulla K+, C+ yellowish, P− (unidentified substances). Common on hardwood trees in open woods and along roads. This species can be recognized in the field by scratching off the cortex with a fingernail to reveal the pigmented medulla.

136b Medulla pale yellow; lobe margins entire, not splitting. .... *Physconia enteroxantha* (see *P. detersa*, page 119)

137a (108) Thallus isidiate (use hand lens and compare with Fig. 155); apothecia rare. 138

137b Thallus without isidia, the surface smooth to wrinkled (lobulate in *Phaeophyscia imbricata*); apothecia common. 154

138a (137) Medulla pale yellow (section with razor blade as shown in Fig. 18). Fig. 156. ...... *Parmelina obsessa* (Ach.) Hale

Figure 156

Figure 156   *Parmelina obsessa*

Thallus greenish mineral gray, closely adnate, 3-6 cm broad; upper surface becoming densely isidiate; margins with very short cilia, especially in the axils; lower surface densely rhizinate; apothecia rare. Cortex K+ yellow

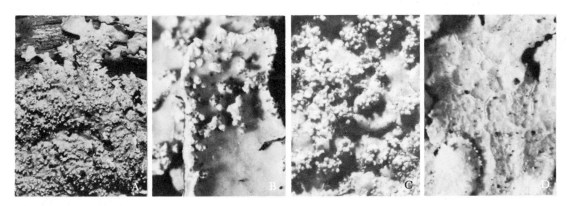

Figure 155

Figure 155   Types of isidia in *Parmelia obsessa* (A, ×5), *Parmelia saxatilis* (B, ×10), *Dirinaria aspera* (C, ×10), and *Pseudoparmelia caroliniana* (D, ×5)

(atranorin); medulla K+, C+, KC+ yellow, P+ orange (galbinic acid and terpenes). Common on large outcrops in open woods. This lichen may be difficult to collect because it prefers broad flat surfaces of hard rocks.

138b Medulla white. 139

139a (138) Thallus loosely attached, the lobes suberect (as in Fig. 3C), lacking rhizines. 140

139b Thallus closely attached with adnate lobes (see Fig. 3A); rhizines or tomentum present, often richly developed (lacking only in *Dirinaria aspera*). .. 141

140a (139) Collected in western North America. Fig. 157. .... *Platismatia herrei* (Imsh.) Culb. & Culb.

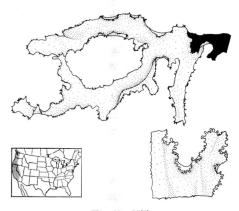

Figure 157

Figure 157 Platismatia herrei (×2)

Thallus greenish mineral gray, loosely attached, 4-8 cm broad; lobes often slightly convoluted; isidia round to flattened and branched, concentrated along the margins; lower surface variable, nearly white to brown or black or mottled; apothecia lacking. Cortex

K+ yellow (atranorin); medulla K−, C−, P− (fatty acids). Common on trunks and branches of conifers. This lichen has much narrower lobes than *Platismatia glauca. P. stenophylla* is similar except in lacking isidia.

140b Collected in eastern North America. Fig. 158. .... *Pseudevernia consocians* (Vain.) Hale & Culb.

Figure 158

Figure 158 Pseudevernia consocians

Thallus subfruticose, light mineral gray, loosely attached; 5-10 cm broad; upper surface moderately to densely isidiate; lower surface white to mottled black; apothecia very rare. Cortex K+ yellow (atranorin); medulla C+, KC+ red (lecanoric acid). Common on conifers in mountainous areas. In the Rocky Mountains all collections of this genus are *Pseudevernia intensa*, which lacks isidia and has abundant apothecia. Both of these species had previously been called *Parmelia furfuracea*, a similar European species with physodic or olivetoric acid.

141a (139) Tips of lobes with angular white markings (use low power lens and compare with Fig. 137). 142

141b Tips of lobes continuous, without white markings, smooth or cracked (round pores only in *Parmelia rudecta*). 143

142a (141) Rhizines densely branched (use hand lens); collected in eastern North America. Fig. 159. ....
.... *Parmelia squarrosa* **Hale**

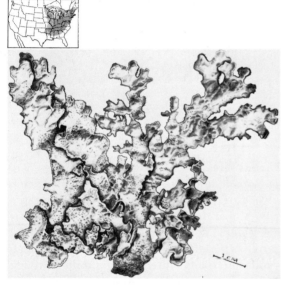

Figure 159

Figure 159   Parmelia squarrosa (×2)

Thallus whitish mineral gray, loosely adnate on bark, up to 12 cm broad; lobes linear, 2-5 mm wide, the surface with distinct white markings toward the tips, cracked, with dense, irregularly inflated isidia; lower surface jet black and densely rhizinate; apothecia rare. Cortex K+ yellow (atranorin); medulla K+ yellow turning red, C−, P+ orange (salazinic acid). Very common on tree trunks, rarely on rocks, in open

forests and swamps. The squarrose rhizines (see Fig. 503) separate this lichen from *P. saxatilis,* which has simple rhizines. It was formerly lumped with *P. saxatilis.*

142b Rhizines simple and unbranched; collected in western North America, at high elevations (above 4000 ft.), or along exposed lake shores in eastern North America. Fig. 160. ....
.... *Parmelia saxatilis* **(L.) Ach.**

Figure 160

Figure 160   Parmelia saxatilis

Thallus whitish to greenish or brownish gray, adnate, 4-15 cm broad; isidia coarse to granular, occurring mostly on ridges and the margin; lower surface moderately rhizinate; apothecia not common. Cortex K+ yellow (atranorin); medulla K+ yellow turning red, C−, P+ orange (salazinic acid often accompanied by lobaric acid). Widespread and common on trees and boulders in open woods or talus slopes. *Parmelia pseudosulcata* Gyel. from Oregon and Washington is identical except for chemistry (K−, P+ red, protocetraric acid).

143a (141) Lower surface densely tomentose (use hand lens and compare with Fig. 7C). Fig. 161. .... *Coccocarpia cronia* (Tuck.) Vain.

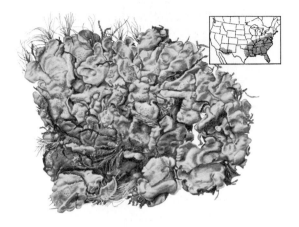

Figure 161

Figure 161   Coccocarpia cronia (×1.5)

Thallus adnate, 1-4 cm broad; lobes spreading, to 5 mm wide; lower surface variable, gray to blackening, the tomentum matted; apothecia very rare. Cortex K−; medulla K−, C−, P− (no substances). Widespread at the base of deciduous trees and on shaded rocks. The non-isidiate relative is *Coccocarpia erythroxyli*.

143b Lower surface rhizinate, shiny, corticate areas easily seen between the rhizines (rhizines lacking only in *Dirinaria aspera*). 144

144a (143) Upper surface with distinct white pores (use hand lens and compare with Fig. 85); lower surface uniformly brown. .... (p. 58) *Parmelia rudecta*

144b Upper surface without pores, continuous or cracked; lower surface black at the center, brown or black at the margins. 145

145a (144) Upper surface finely cracked to the margin (use hand lens and compare with Fig. 155D); lobes irregularly broadened. Fig. 162. .... *Pseudoparmelia caroliniana* (Nyl.) Hale

Figure 162

Figure 162   Pseudoparmelia caroliniana

Thallus greenish mineral gray, adnate, 5-10 cm broad; upper surface moderately isidiate; lower surface black or dark brown, naked in a narrow marginal zone; apothecia rare. Cortex K+ yellow (atranorin); medulla KC+ faint purple (perlatolic acid). Common in open woods, often on conifers, more rarely on rocks. It resembles *Parmelia rudecta* which differs in being C+ red and in having a very pale lower surface.

145b Upper surface continuous, irregularly cracked only on older lobes; lobes generally narrow and linear. 146

146a (145) Marginal cilia basally inflated (use hand lens; do not confuse projecting rhizines in *Hypotrachyna* species for cilia). 147

146b Marginal cilia simple or absent. **148**

147a (146) Medulla C+ deep red; lower surface jet black. Fig. 163. ....
.... *Bulbothrix laevigatula* (Nyl.) Hale

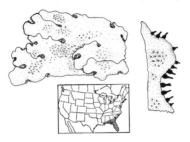

Figure 163

Figure 163   Bulbothrix laevigatula (×5)

Thallus whitish gray, closely adnate, 3-8 cm broad; isidia short, sparse to quite dense; lower surface densely rhizinate, the rhizines branched; apothecia not common. Cortex K+ yellow (atranorin); medulla K−, C+ red, P− (lecanoric acid). On branches and trunks of deciduous trees in open woods. This species is closely related to *B. goebelii* (below), which has a dark brown lower surface and different chemistry.

147b Medulla C+ rose; lower surface dark brown. Fig. 164. ....
.... *Bulbothrix goebelii* (Zenk.) Hale

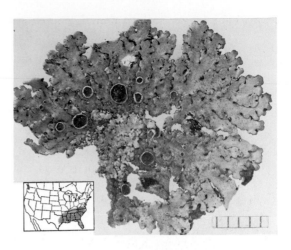

Figure 164

Figure 164   Bulbothrix goebelii

Thallus whitish to greenish mineral gray (often turning buff in the herbarium), fragile, closely adnate, 3-6 cm broad; lobes about 1 mm wide, becoming dissected marginally; lower surface brown or darkening; apothecia rare. Cortex K+ yellow (atranorin); medulla K−, C+ rose, P− (gyrophoric acid). Rather widespread but inconspicuous on trunks and branches in open woods.

148a (146) Rhizines richly dichotomously branched (use hand lens and see Fig. 484), often projecting as a mat around the margin. **149**

148b Rhizines simple or furcate, usually not projecting beyond the margin (lacking in *Dirinaria aspera*). **151**

149a (148) Isidia inflated, hollow, pustulate and fragile (use hand lens and compare with Fig. 136). ....
.... (p. 82) *Hypotrachyna formosana*

149b Isidia thin, solid, firm. **150**

150a (149) Isidia cylindrical, produced over the whole upper surface; medulla C— or C+ pale orange. Fig. 165. .... *Hypotrachyna imbricatula* (Zahlbr.) Hale

Figure 165

Figure 165   Hypotrachyna imbricatula

Thallus light mineral gray (turning buff in the herbarium), adnate, 5-10 cm broad; upper surface sparsely to moderately isidiate; apothecia lacking. Cortex K+ yellow (atranorin); medulla K—, C— or C+ orange, KC+ orange, P— (barbatic acid group). On tree trunks in open woods in the Southern Appalachians. *Hypotrachyna dentella* (Hale & Kurok.) Hale, rarely collected in the same range, differs in having a P+ red reaction (echinocarpic acid).

150b Isidia becoming flattened (see Fig. 13) and in part produced on margins and tips of lobes; medulla C+ red. Fig. 166. .... *Hypotrachyna prolongata* (Kurok.) Hale

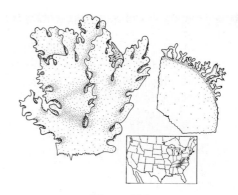

Figure 166

Figure 166   Hypotrachyna prolongata (×2)

Thallus light mineral gray, adnate, 5-10 cm broad; lower surface densely rhizinate, the rhizines dichotomously branched; apothecia lacking. Cortex K+ yellow (atranorin); medulla C+, KC+ red (anziaic acid). On bark of spruce and fir at high elevations in the Great Smoky Mountains in North Carolina and in southern Virginia. This often occurs with *H. rockii*. Another associated species, *H. ensifolia* (Kurok.) Hale, is very close but reacts C—, KC+ rose (alectoronic acid).

151a (148) Isidia coarse and subsorediate, breaking open (use hand lens and compare with Fig. 155), crater-shaped. Fig. 167. ....
.... *Dirinaria aspera* (Magn.) Awas.

Figure 167

Figure 167 Dirinaria aspera (×4)

Thallus whitish gray, closely adnate on bark, 2-5 cm broad; lobes crowded to confluent, to 1 mm wide; upper surface becoming densely isidiate; lower surface black, lacking rhizines; apothecia lacking. Cortex K+ yellow (atranorin); medulla K−, C−, P− (divaricatic acid). Common on oaks and other hardwoods in open forests and pastures. This species might be misidentified as *Pyxine eschweileri,* which has distinct rhizines and powdery soredia.

151b Isidia fine, cylindrical or flattened, not breaking open. **152**

152a (151) Isidia in part flattened and apically ciliate (use hand lens). Fig. 168. .... **Parmelina horrescens (Tayl.) Hale**

Figure 168

Figure 168 Parmelina horrescens (×10)

Thallus light mineral gray, adnate, 2-4 cm broad; isidia moderate to dense; lower surface sparsely to moderately rhizinate; apothecia rare. Cortex K+ yellow (atranorin); medulla K−, C−, KC+ rose, P− (unknown substances). Fairly widespread on conifers and hardwoods in open woods, often occurring with *P. dissecta.*

152b Isidia cylindrical, without apical cilia. **153**

153a (152) Marginal cilia present; medulla C+ rose. Fig. 169. .... **Parmelina dissecta (Nyl.) Hale**

Figure 169

Figure 169 Parmelina dissecta

Thallus light mineral gray, adnate, 3-7 cm broad; upper surface moderately isidiate; lower surface moderately rhizinate, the rhizines simple; apothecia rare. Cortex K+ yellow (atranorin); medulla C+, KC+ red (gyrophoric acid). Common on exposed trees or rocks in dry areas. *Parmelia horrescens* (above) often occurs with this species and must be carefully distinguished from it with the C test. Specimens with a yellow medulla are probably *P. obsessa.*

153b Marginal cilia lacking; medulla C−. Fig. 170. ....
Pseudoparmelia amazonica (Nyl.) Hale

Figure 170

Figure 170   Pseudoparmelia amazonica

Thallus whitish mineral gray (often turning buff in the herbarium), adnate on bark, 3-6 cm broad; lobes irregularly broadened, 2-5 mm wide; apothecia lacking. Cortex K+ yellow (atranorin); medulla K−, C−, P+ red (protocetraric acid). On oaks and palm trees in open woods, often with *P. salacinifera.* Three superficially similar species may key out here: *P. martinicana* (Nyl.) Hale (medulla K−, P+ red, protocetraric acid), which has coarse isidia and has been collected twice in Florida; *Parmotrema crinitum* (medulla K+ yellow, stictic acid), a widespread species, sometimes with rhizines produced to the margin; and *Parmotrema internexum* (Nyl.) Hale (medulla K+ yellow, stictic acid), smaller than *P. crinitum,* but with short marginal cilia.

154a (137) Lower surface with a thick spongy layer of tomentum (use hand lens and compare with Fig. 171). **155**

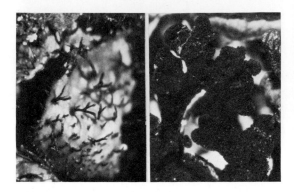

Figure 171

Figure 171   Rhizines of *Hypotrachyna livida* (left) and spongy tomentum of *Anzia colpodes* (about ×10)

154b Lower surface corticate, moderately to sparsely rhizinate or bare. **156**

155a (154) Lobes narrow and linear, 1-2 mm wide, appearing inflated. Fig. 172. ....
.... *Anzia colpodes* (Ach.) Stizb.

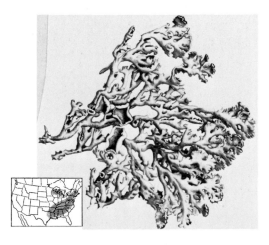

Figure 172

Figure 172   Anzia colpodes (×1)

Thallus light mineral gray, leathery, adnate, 3-7 cm broad; pycnidia common on upper surface; apothecia common, spores tiny (1 micron), numerous in each ascus. Cortex K+ yellow (atranorin); medulla K−, C−, P− (divaricatic acid). Widespread on deciduous trees, especially oak and hickory, but not commonly collected. This species is probably overlooked because it often grows high up on tree trunks. Superficially it looks like a *Hypogymnia* but the lobes are solid.

155b Lobes broader, 2-4 mm wide, flat. Fig. 173. .... *Coccocarpia erythroxyli* (Spreng.) Swinsc. & Krog

Figure 173

Figure 173   Coccocarpia erythroxyli (×1.5)

Thallus dark mineral gray, adnate, 3-6 cm broad; lower surface black to deep gray, densely tomentose; apothecia common, black, adnate on the thallus. Cortex K−; medulla K−, C−, P− (no substances). Widespread on deciduous trees in open woods and along roads. This species, closely related to isidiate *Coccocarpia cronia,* is most common in Florida.

156a (154) Thallus adnate to suberect, almost fruticose (see Fig. 3C); rhizines lacking. **157**

156b Thallus adnate to closely appressed (see Fig. 3A); rhizines usually present (lacking only in *Cavernularia lophyrea, Dirinaria confusa,* and *Hypogymnia oroarctica*). **160**

157a (156) Lower surface flat; medulla C−. **158**

157b Lower surface channelled (see Fig. 481); medulla C+ red. **159**

158a (157) Lower surface jet black, wrinkled (use lens). Fig. 174. .... *Cetraria idahoensis* Essl.

Figure 174

Figure 174   Cetraria idahoensis

Thallus greenish mineral gray, loosely attached to suberect on branches, leathery and firm, 4-7 cm broad; lobes 1-4 mm wide, deeply dissected and black rimmed, with numerous pycnidia (visible with hand lens); lower surface deeply wrinkled, black to dark brown, very sparsely rhizinate; apothecia common, short-stalked, the disc brown. Cortex K+ yellow (atranorin); medulla K−, C−, P− (fatty acids). Common on branches of exposed conifers or in canopy of trees in dense forests. Once recognized, this *Cetraria* cannot be confused with any other species.

158b Lower surface black at the center and mottled to brown and white at the tips, shiny and smooth. Fig. 175. .... *Platismatia stenophylla* (Tuck.) Culb. & Culb.

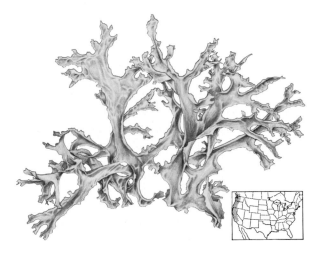

Figure 175

Figure 175   Platismatia stenophylla (×2)

Thallus light mineral gray, often browning at the tips, 6-12 cm broad; upper surface smooth to wrinkled; apothecia not common. Cortex K+ yellow (atranorin); medulla K−, C−, P− (fatty acids). Common on exposed conifers. *Platismatia herrei* is close but has marginal isidia; *P. glauca* (see page 63) has much broader lobes and soredia.

159a (157) Collected in eastern North America. Fig. 176. .... *Pseudevernia cladonia* (Tuck.) Hale & Culb.

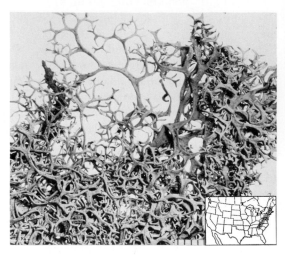

Figure 176

Figure 176   Pseudevernia cladonia

Thallus fruticose, suberect on twigs, light mineral gray, 4-10 cm broad; branches somewhat dorsiventral; lower surface white to mottled black, naked; apothecia very rare. Cortex K+ yellow (atranorin); medulla C+ red, KC+ red (lecanoric acid). Common on conifers at high elevations (4-6000 ft.) *Pseudevernia consocians* commonly occurs with it but has isidia and much wider branches.

159b Collected in western North America. Fig. 177. .... *Pseudevernia intensa* (Nyl.) Hale & Culb.

Figure 177

Figure 177 Pseudevernia intensa (×1)

Thallus light mineral gray, loosely attached on bark, 5-10 cm broad; upper surface smooth to deeply wrinkled, black pycnidia common; lower surface black at the center but usually turning mottled buff to white at the margin; apothecia common. Cortex K+ yellow (atranorin); medulla C+, KC+ red (lecanoric acid). Common on exposed conifers.

160a (156) Upper cortex with reticulate white markings, especially toward the margins, plane to weakly ridged (use hand lens and compare with Fig. 6C). Fig. 178. .... *Parmelia omphalodes* (L.) Ach.

Figure 178

Figure 178 Parmelia omphalodes (×1)

Thallus greenish or whitish mineral gray to brown, loosely adnate, 6-15 cm broad; lower surface black and moderately rhizinate; apothecia common. Cortex K+ yellow (atranorin); medulla K+ yellow→red, P+ orange (salazinic acid). Widespread on exposed boulders in open areas and talus slopes. This is a common lichen on rocks in arctic and alpine localities and could only be confused with *Parmelia saxatilis* (see page 89), which has coarse isidia.

160b Upper cortex continuous, plane to wrinkled, shiny, or pruinose. 161

161a (160) Upper cortex scabrid, becoming white pruinose (use hand lens and compare with Fig. 6D); cortex K−. .... (p. 144) *Physconia muscigena* and (p. 138) *P. pulverulenta*

161b Upper cortex shiny, rarely becoming white pruinose; cortex K+ yellow (except K− in *Phaeophyscia* species). .. 162

162a (161) Collected in western North America. 163

162b Collected in eastern and/or southern United States (and adjacent Canada). 169

163a (162) Collected on trees. 164

163b Collected on rocks or mosses over rocks. 168

164a (163) Lower surface moderately to densely rhizinate (use hand lens). 165

164b Lower surface bare, smooth or wrinkled. 167

165a (164) Lobes 0.5-1 mm wide (use ruler); cortex K−. .... (p. 140) *Phaeophyscia ciliata*

165b Lobes 2-5 mm wide; cortex K+ yellow. 166

166a (165) Rhizines simple to sparsely branched; medulla C+ red. Fig. 179. .... *Parmelina quercina* (Willd.) Hale

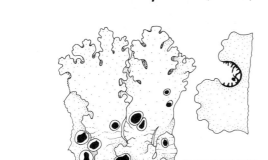

Figure 179

Figure 179 Parmelina quercina (×2)

Thallus light mineral gray, adnate, 3-10 cm broad; lower surface black and densely rhizinate, with some cilia in lobe axils; apothecia common. Cortex K+ yellow (atranorin); medulla C+, KC+ red (lecanoric acid). Common on oak trees. Reports of this species from eastern North America have been based on *Hypotrachyna livida* or *Parmelina galbina*.

166b Rhizines richly dichotomously branched (see Fig. 483); medulla C+ rose. .... *Hypotrachyna pulvinata* (Fée) Hale (see *H. livida*, page 101)

167a (164) Lobes 1-5 mm wide, irregularly broadened. .... *Cetraria idahoensis* (see page 95)

167b Lobes narrow and linear, about 1 mm wide. Fig. 180. .... *Cavernularia lophyrea* (Ach.) Degel.

Figure 180

Figure 180 Cavernularia lophyrea

Thallus greenish mineral gray (sometimes turning buff in the herbarium), adnate, 2-4 cm broad; lobes 1-1.5 mm wide, branched and dissected; upper surface shiny, pycnidiate; lower surface shiny, dark brown to blackening, perforated with small holes or pits; apothecia common. Cortex K+ yellow (atranorin); medulla K−, C−, P− (physodic acid). On conifers and fence posts, rather rarely collected and probably overlooked. This unusual lichen differs from *Hypogymnia* in having a solid medulla. The pores below open into invaginated chambers with a continuous cortex.

168a (163) Rhizines lacking (use hand lens); lobes appearing inflated. Fig. 181. .... *Hypogymnia oroarctica* **Krog**

Figure 181

Figure 181   Hypogymnia oroarctica

Thallus brownish mineral gray, closely adnate on rock but sometimes subascending marginally, 4-7 cm broad; lobes inflated but solid, elongate and little branched, 0.5-1 mm wide; lower surface smooth to wrinkled, black; apothecia rare. Upper cortex K+ yellow (atranorin); medulla K−, C−, P− (physodic acid). Common in arctic-alpine areas. This arctic lichen is related to the corticolous Hypogymnias of the conifer forests but has solid lobes.

168b Rhizines present; lobes flat to convex, dissected. .... **(p. 99)** *Phaeophyscia decolor*

169a (162) Lobes irregularly broadened, 2-6 mm wide, white pores visible with hand lens toward lobe tips; upper surface becoming densely lobulate. .... **(p. 58)** *Parmelia appalachensis*

169b Lobes narrow and more linear 0.5-3 mm wide, lacking pores (lobulate only in *Phaeophyscia imbricata*). **170**

170a (169) Lower surface lacking any rhizines (use hand lens); lobes appressed, confluent and crowded. Fig. 182. .... *Dirinaria confusa* **Awas.**

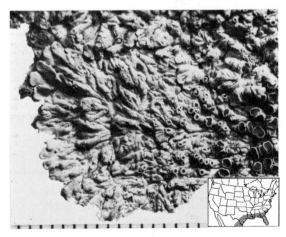

Figure 182

Figure 182   Dirinaria confusa

Thallus whitish mineral gray, 4-10 cm broad, sometimes covering extensive areas of bark; apothecia very common, the disc black. Cortex K+ yellow (atranorin); medulla K−, C−, P− (sekikaic acid). Common on exposed trees in the Gulf states, rarely also on rocks. *Dirinaria purpurascens* (Vain.) Moore is extremely close but has a purplish-red pruinose disc. It is known mostly from Louisiana and Florida. Species of *Pyxine* which key out here can be recognized by the presence of simple rhizines below.

170b Lower surface clearly rhizinate; lobes crowded but usually separate. 171

171a (170) Margins of lobes densely lobulate (use hand lens and see Fig. 269). .... (p. 141) *Phaeophyscia imbricata*

171b Margins of lobes not lobulate. 172

172a (171) Margins of lobes with dense, basally inflated cilia (use hand lens). Fig. 183. .... *Bulbothrix confoederata* (Culb.) Hale

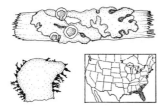

Figure 183

Figure 183   Bulbothrix confoederata (×5)

Thallus whitish mineral gray, appressed, 2-5 cm broad; lower surface black, densely rhizinate, the rhizines branched; apothecia common, the rim covered with black dots (pycnidia). Cortex K+ yellow (atranorin); medulla C+, KC+ red (lecanoric acid). Common in exposed scrub forests on twigs and branches. This species is easily overlooked because of the small size. Another species with inflated cilia, *Bulbothrix coronata* (Fée), Hale, reacts C+ rose (gyrophoric acid) and has been collected in western Texas.

172b Margins of lobes lacking inflated cilia or eciliate. **173**

173a (172) Thallus brownish mineral gray; cortex K−. .... (p. 140) *Phaeophyscia ciliata*

173b Thallus light greenish to whitish mineral gray; cortex K+ yellow. **174**

174a (173) Rhizines simple and unbranched (use hand lens); lobes with short cilia in the axils; medulla often pale orange under apothecia. Fig. 184. .... *Parmelina galbina* (Ach.) Hale

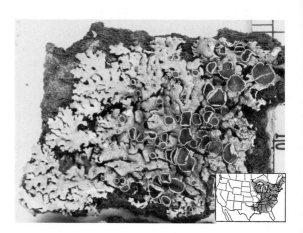

Figure 184

Figure 184   Parmelina galbina

Thallus greenish mineral gray, closely adnate, 3-10 cm broad; upper surface becoming finely wrinkled with numerous small black dots (pycnidia); margins short-ciliate, especially in the axils; medulla pale yellow orange, at least beneath apothecia; apothecia very common. Cortex K+ yellow (atranorin); medulla K+ yellow→red, P+ orange (galbinic acid). Very common on trunks and branches of deciduous trees in open forests. *Hypotrachyna livida* (below) has often been confused with this species but has different rhizines and chemistry. Both were called *Parmelia quercina* or *P. sublaevigata* in the past.

174b Rhizines sparsely dichotomously branched (use hand lens and see Fig. 484); lobes without cilia; medulla white. Fig. 185. .... *Hypotrachyna livida* (Tayl.) Hale

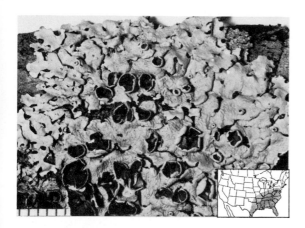

Figure 185

Figure 185 *Hypotrachyna livida*

Thallus whitish mineral gray, closely adnate, 4-9 cm broad; lower surface moderately rhizinate; apothecia very common. Cortex K+ yellow (atranorin); medulla KC+ rose (4-0-methyl physodic acid and lividic acid). Common on trunks and branches of deciduous trees in open woods or along roadsides. An almost identical species, *Hypotrachyna pulvinata* (Fée) Hale, occurs on trees in southern Arizona and New Mexico; it contains evernic acid. Another species in this genus, *H. virginica* (Hale) Hale, will be rarely collected in the southern Appalachian Mountains; it is C+, KC+ orange (barbatic acid group) and has a roughened, subpustulate surface.

NARROW-LOBED MINERAL GRAY LICHENS WITH A TAN OR WHITE LOWER SURFACE

175a (97) Thallus with soredia on the tips or upper surface of lobe (use lens and see Fig. 186); apothecia rarely present. .. 176

Figure 186

Figure 186 Soredia of *Physcia tribacoides* (left) and *P. millegrana* (×10)

175b Thallus lacking soredia, either isidiate or lacking soredia and isidia; apothecia frequently present. 192

176a (175) Lobes rather finely branched and narrow, 0.1-0.3 mm wide (use ruler). 177

176b Lobes broader, 0.5-3 mm wide, not finely branched. 178

177a (176) Collected on trees. Fig. 187. .... *Physcia millegrana* Degel.

Figure 187

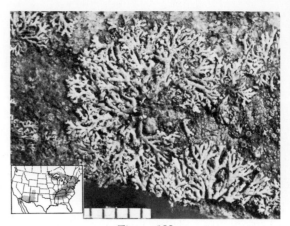

Figure 188

Figure 187 Physcia millegrana

Thallus whitish mineral gray, adnate to closely attached, 1-2 cm broad but fusing into larger colonies; lower surface white, sparsely rhizinate; apothecia sometimes present. Cortex K+ yellow (atranorin). Very common on roadside trees and in open woods and rarely also on rocks. This weedy species seems to withstand pollution near cities better than most lichens. Specimens reported from California were probably introduced on nursery tree stock from the East.

177b Collected on rocks. Fig. 188. ....
.... *Physcia subtilis* **Degel.**

Figure 188 Physcia subtilis

Thallus whitish mineral gray, closely adnate, 0.5-1.0 cm broad, often coalescing into larger colonies; lower surface white, sparsely rhizinate; apothecia rare. Cortex and medulla K+ yellow (atranorin). Widespread on exposed granitic boulders or sandstone. *Physcia halei* is similar but lacks soredia. Another saxicolous species, *P. dubia,* is larger (lobes more than 0.3 mm wide) with larger laminal or apical soralia.

178a (176) Margins of lobes ciliate (use hand lens and do not confuse a few projecting rhizines as cilia, especially in *Heterodermia*). 179

178b Margins of lobes without cilia but occasionally with some projecting rhizines. 180

179a (178) Cilia very long, 2-4 mm; lobes long and almost subfruticose; soralia on the lower surface or tips. Fig. 189. ....
.. *Heterodermia leucomelaena* (**L.**) **Poelt**

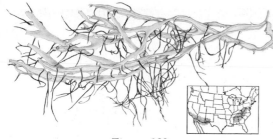

Figure 189

Figure 189 Heterodermia leucomelaena

Thallus light mineral gray, loosely attached, 2-8 cm broad; lower surface white and cottony, cortex lacking; apothecia rare. Upper cortex K+ yellow (atranorin); medulla K+ yellow→ red, P+ pale orange (salazinic acid). Rather rare on deciduous trees in open woods. In the Appalachian Mountains *H. appalachensis* (Kurok.) Culb. is nearly as common; it is K— in the medulla and has a pale yellow lower surface.

179b Cilia shorter, 1-2 mm long; lobes short and crowded, subascending; soralia apical in hooded tips. Fig. 190. .... *Physcia adscendens* (Fr.) Oliv.

Figure 190

Figure 190 Physcia adscendens

Thallus light mineral gray, adnate, 2-4 cm broad; soredia under the lobe tips; marginal cilia conspicuous; lower surface white, moderately rhizinate; apothecia very rare. Cortex K+ yellow (atranorin). Common on deciduous trees and conifers in exposed areas and along roadsides. This unusual lichen is characterized by the hood-shaped lobe tips. *Physcia tenella* (Scop.) DC., a related ciliate species, has flattened lobe tips, also sorediate beneath. It will be found most often in western North America, often on rocks.

180a (178) Lower surface fibrous or cottony, white or pale orange (use hand lens and see Fig. 191); rhizines lacking except at the margins. 181

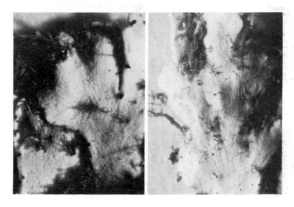

Figure 191

Figure 191 Fibrous ecorticate lower surface of *Heterodermia hypoleuca* (left) and corticate surface of *H. speciosa* (×10)

180b Lower surface smooth, corticate and shiny, creamy white to buff or light brown (see Fig. 191). 183

181a (180) Lower surface purplish black toward the center. Fig. 192. .... *Heterodermia casarettiana* (Mass.) Trev.

Figure 192

Figure 192   *Heterodermia casarettiana* (×2)

Thallus whitish gray, loosely adnate, brittle, 4-8 cm broad; lobes long and linear, 1-2 mm wide; soredia apical, in capitate to lip-shaped soralia; lower surface white (or pigmented pale yellow) toward the tips with long, black, squarrosely branched rhizines along the margin; apothecia rare. Cortex K+ yellow; (atranorin); medulla K+ yelow turning red, C−, P+ orange (norstictic acid) or K−, P− (no substances). Common at the base of trees and on rocks in open woods. This, *H. hypoleuca,* and *H. squamulosa* are the only Heterodermias with a dark lower surface. This whole group, however, needs further study, as some specimens will not seem to fit in the keys.

181b Lower surface uniformly white or with patches of orange. 182

182a (181) Lower surface in part orange (use hand lens and examine lobe tips). Fig. 193. .... *Heterodermia obscurata* (Nyl.) Trev.

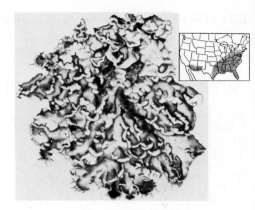

Figure 193

Figure 193   *Heterodermia obscurata* (×1.5)

Thallus light mineral gray, adnate, 3-8 cm broad; lower surface orange and cottony, cortex lacking; rhizines marginal; apothecia very rare. Cortex K+ yellow (atranorin). Common on deciduous trees, more rarely on rocks, in open woods and along roadsides. It is very close to *H. speciosa* (below), which is white below. *H. casarettiana* may have a yellowish pigment below but it reacts K− in contrast to the K+ purple of *H. obscurata.*

182b Lower surface white. ....
.... (p. 105) *Heterodermia speciosa*

Figure 194

Figure 194   Schematic comparison of marginal and laminal soralia

183a (180) Soredia marginal in linear soralia (see Fig. 194A). Fig. 195. .... *Heterodermia albicans* (Pers.) Swinsc. & Krog

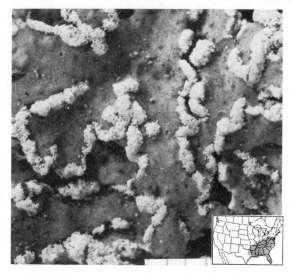

Figure 195

Figure 195 Heterodermia albicans

Thallus light mineral gray, adnate, 4-6 cm broad; margins with some short projecting rhizines; lower surface white to buff, moderately rhizinate; apothecia rare. Cortex K+ yellow (atranorin); medulla K+ yellow→red, P+ pale orange (salazinic acid). Fairly common on roadside trees and in open woods. *Heterodermia speciosa* has mostly apical crescent-shaped soralia and a K+ yellow medulla. A similar marginally sorediate species, *Physcia crispa* (see page 79), has a dark lower surface and a K+ yellow medulla.

183b Soredia laminal or apical in round or labriform soralia (see Fig. 194). 184

184a (183) Lobes irregularly broadened, 2-5 mm wide (use ruler); soredia originating from white pores (pseudocyphellae); medulla C+ red. .... (p. 56) *Parmelia subrudecta*

184b Lobes narrow and linear, 0.5-2 mm wide; pseudocyphellae absent; medulla C−. 185

185a (184) Soralia strongly apical in irregular to labriform soralia (*Physcia dubia* may have apical soralia but lobes are less than 1 mm wide). 186

185b Soralia laminal or in part marginal. .. 187

186a (185) Soralia strongly crescent-shaped, the soredia abundant and powdery. Fig. 196. .... *Heterodermia speciosa* (Wulf.) Trev.

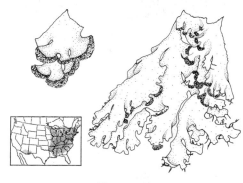

Figure 196

Figure 196 Heterodermia speciosa (×2)

Thallus whitish mineral gray, adnate, 2-8 cm broad; lower cortex entire but sometimes eroding away and appearing ecorticate, sparsely rhizinate; apothecia rare. Cortex and medulla K+ yellow (atranorin and zeorin) or K+ yellow turning red (norstictic acid). Common on deciduous trees and mossy rocks in mature woods and swamps. The norstictic acid population has been called *H. pseudospeciosa* (Kurok.) Culb. Other confusable species with labriform apical soralia include *H. obscurata* which is ecorticate and orange below, *H. casarettiana,* more robust, ecorticate, and blackening below, and *Physcia pseudospeciosa* (below) with sparse soredia and a paraplectenchymatous cortex (rather than prosoplectenchymatous as in *Heterodermia*) as seen in a microscopic section.

186b Soralia mostly irregular, not strongly crescent-shaped, the soredia rather sparse and granular. Fig. 197. .... *Physcia pseudospeciosa* **Thoms.**

Figure 197

Figure 197 Physcia pseudospeciosa

Thallus whitish gray, rather loosely adnate over mosses on rocks, 3-8 cm broad; lobes rather long and narrow, 1-1.5 mm wide, the upper surface faintly white-spotted (as in *P. aipolia*); soralia terminal and marginal on the lower surface, lip-shaped with lobe tips turning up; apothecia lacking. Cortex and medulla K+

yellow (atranorin and zeorin). Widespread, but not commonly collected, on mossy rocks in open oak forests in the southern Appalachian region. Closely related *P. caesia* has shorter lobes and round laminal soralia.

187a (185) Collected on trees. **188**

187b Collected on rocks. **190**

188a (187) Thallus greenish to brownish gray; cortex K−; lobes less than 1 mm wide. .. .. **(p. 122)** *Physcia chloantha*

188b Thallus whitish gray; cortex K+ yellow; lobes 1-2 mm wide. **189**

189a (188) Medulla K+ yellow; collected on deciduous trees in the eastern forests. Fig. 198. .... *Physcia americana* **Merr.**

Figure 198

Figure 198 Physcia americana

Thallus whitish gray, adnate closely on bark, 1-3 cm broad; upper cortex shiny, becoming faintly white-spotted (use lens and see Fig.

6E); lower surface white, moderately rhizinate; apothecia rare. Cortex and medulla K+ yellow (atranorin and an unknown terpene). Common in open deciduous forests, especially on white oak and ash. Rare collections of *P. caesia* on bark can be separated by the strong white-spotting. In California this species is replaced by *P. clementi* (Sm.) Lynge, which has more diffuse soralia. Specimens of *Dirinaria picta* that key out here can be distinguished by the black lower surface.

189b Medulla K−; collected on conifers and *Populus* **and other northern hardwoods in northern and western North America. Fig. 199.** ....
...... *Parmeliopsis hyperopta* (Ach.) Arn.

Figure 199

Figure 199   Parmeliopsis hyperopta

Thallus whitish gray, closely adnate (usually collected with bark), 2-6 cm broad; lobes about 1 mm wide; soredia powdery in round subterminal and laminal soralia; lower surface tan to darker brown, moderately rhizinate; apothecia not common. Cortex K+ yellow (atranorin); medulla K−, C−, P− (divaricatic acid). Common at the base of conifers and hardwoods in northern and western forests. This species is the gray, usnic acid-free counterpart of *P. ambigua* and the two often grow together.

190a (187) Upper cortex white-spotted (use hand lens and compare with Fig. 6E); soralia mostly laminal. Fig. 200. ....
.... *Physcia caesia* (Hoffm.) Hampe

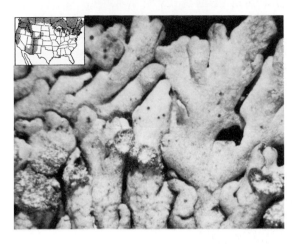

Figure 200

Figure 200   Physcia caesia (×10)

Thallus whitish mineral gray, closely adnate, 4-8 cm broad; lower surface white to buff, moderately rhizinate; apothecia rare. Cortex and medulla K+ yellow (atranorin and zeorin). Widespread on boulders and cliffs in fairly exposed areas. This is apparently the sorediate morph of *P. phaea* (see page 113).

190b Upper cortex uniform, not white-spotted; soralia marginal and apical. 191

191a (190) Lobes 0.3-1 mm wide (use ruler), closely adnate on rock. Fig. 201. ....
.... *Physcia dubia* (Hoffm.) Lett.

Figure 201

Figure 201   Physcia dubia

Thallus whitish mineral gray, closely adnate, 2-5 cm broad; lobes 0.5-1.0 mm wide; lower surface white to buff, moderately rhizinate; sorediate lobe tips often turning up; apothecia rare. Cortex K+ yellow (atranorin); medulla K−, C−, P− (no substances). Widespread on exposed rocks or open talus slopes; difficult to remove from rock and frequently overlooked. The soralia are extremely variable, mostly apical but sometimes laminal.

**191b Lobes 1-2 mm broad, adnate. Fig. 202. ....
.... *Physcia callosa* Nyl.**

Figure 202

Figure 202   Physcia callosa (×2)

Thallus whitish mineral gray, adnate, 2-6 cm broad, the colonies often coalescing; lobe margins becoming finely dissected with scattered soredia; lower surface white, moderately rhizinate; apothecia rare. Cortex and medulla K+ yellow (atranorin). Common on open rock outcrops. There may be some confusion with *P. dubia* which reacts K− in the medulla.

192a (175) Thallus isidiate (use hand lens and compare with Fig. 203); apothecia rare. 193

Figure 203

Figure 203 Isidia of *Parmeliopsis aleurites* and *Heterodermia granulifera* (×10)

192b Thallus lacking isidia, the surface smooth to wrinkled; apothecia commonly developed. 199

193a (192) Lobes irregularly broadened and apically rotund, 2-6 mm wide. 194

193b Lobes generally linear and narrow, 0.5-2 mm wide (see Fig. 99). 196

194a (193) Upper cortex with white pores (examine lobe tips with hand lens and compare with Fig. 85). ....
.... (p. 58) *Parmelia rudecta*

194b Upper cortex without pores, continuous or cracked. 195

195a (194) Upper cortex finely reticulately cracked (use hand lens and compare with Fig. 155); medulla K–. ...... (p. 90) *Pseudoparmelia caroliniana*

195b Upper cortex continuous, cracked only on older lobes; medulla K+ yellow turning red. Fig. 204. .... *Pseudoparmelia salacinifera* (Hale) Hale

Figure 204

Figure 204 Pseudoparmelia salacinifera (×1.5)

Thallus mineral gray, turning buff in the herbarium, adnate, 6-10 cm broad; upper surface moderately to sparsely isidiate; lower surface tan, moderately rhizinate; apothecia lacking. Cortex K+ yellow (atranorin); medulla K+ yellow→red (salazinic acid). On oak and palm trees in Florida. It often occurs with *P. amazonica* which has a black lower surface and K– reaction.

196a (193) Margins of lobes with basally inflated cilia (use hand lens and see Fig. 183); lower surface brown. ....
.... (p. 91) *Bulbothrix goebelii*

196b Margins of lobes lacking cilia; lower surface uniformly white to buff. 197

197a (196) Lower surface channelled (see Fig. 481), lacking rhizines; thallus loosely adnate to subfruticose (as in Fig. 3C). .... (p. 88) *Pseudevernia consocians*

197b Lower surface flat, rhizinate. 198

198a (197) Isidia thin, cylindrical (see Fig. 203); cortex K+ deep yellow. Fig. 205. .. .. *Parmeliopsis aleurites* (Ach.) Nyl.

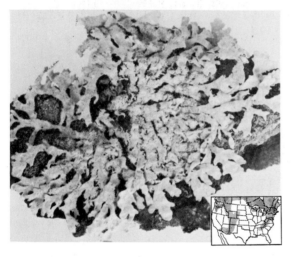

Figure 205

Figure 205 Parmeliopsis aleurites

Thallus whitish mineral gray, adnate, 2-5 cm broad; isidia often becoming dense; lower surface sparsely rhizinate; apothecia very rare. Cortex K+ deep yellow (thamnolic acid). Common on bark of conifers and on dead stumps in open woods, rarely on rocks.

198b Isidia barrel-shaped, becoming granular; cortex K+ light yellow. Fig. 206. .... *Heterodermia granulifera* (Ach.) W. Culb.

Figure 206

Figure 206 Heterodermia granulifera

Thallus light mineral gray, adnate, 3-6 cm broad; isidia moderate, constricted at the base; lower surface moderately rhizinate; apothecia rare. Cortex K+ yellow (atranorin); medulla K+ yellow→red, P+ orange (salazinic acid). On deciduous trees in open woods.

199a (192) Lower surface white and fibrous or cottony (use hand lens and compare with Fig. 191), lacking a cortex; rhizines marginal. 200

199b Lower surface whitish tan to pale brown; cortex and rhizines (or tomentum) present. 202

200a (199) Thallus 2-3 cm broad, the lobes suberect, long ciliate. Fig. 207. .... *Heterodermia echinata* (Tayl.) W. Culb.

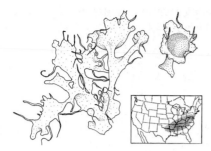

Figure 207

Figure 207   Heterodermia echinata (×2)

Thallus tufted and suberect, 2-3 cm broad; lower surface chalky white, the cortex lacking; apothecia numerous, the rim ciliate. Cortex K+ yellow (atranorin). Locally abundant on juniper twigs and other trees in open pastures. In southern California one will collect a closely related species, *H. erinacea* (Ach.) Hale, which has narrower lobes and grows on shrubs, cacti, and rocks on the coast where fog comes in.

200b Thallus 3-10 cm broad; lobes adnate, without cilia. 201

201a (200) Upper surface and margins with dense erect lobules (Fig. 5B). Fig. 208. .... *Heterodermia squamulosa* (Degel.) W. Culb.

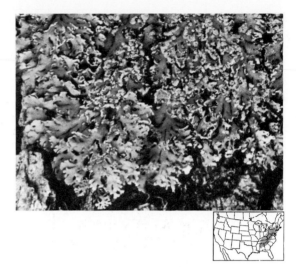

Figure 208

Figure 208   Heterodermia squamulosa (×1.5)

Thallus whitish gray, fragile, adnate on bark, 5-15 cm broad; lobes linear and narrow, 1-1.5 mm wide, the margins and in part surface becoming densely isidiate-lobulate, the lower surface of the upturned lobules white; lower surface white at the tips but darkening toward the center; apothecia lacking. Cortex and medulla K+ yellow (atranorin and zeorin). Common at the base of trees and over mosses on trees in the Appalachian Mountains.

201b Upper surface without erect lobules but sometimes with adnate marginal lobules. Fig. 209. .... *Heterodermia hypoleuca* (Ach.) Trev.

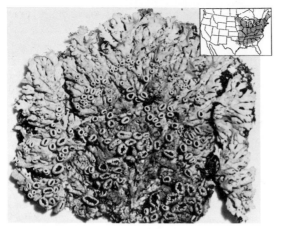

Figure 209

Figure 209   Heterodermia hypoleuca

Thallus whitish mineral gray, adnate, 3-6 cm broad; lobe margins and rim of apothecia becoming sparsely lobulate with age; lower surface with short marginal rhizines; apothecia common. Cortex and medulla K+ yellow (atranorin and zeorin). Widespread on deciduous trees in open woods or along roads. There is some intergradation with *H. squamulosa* (above) but the lobules are appressed.

202a (199) **Upper surface and rim of apothecia with white pores (pseudocyphellae) (use hand lens and compare with Fig. 85); thallus rather large with broadened lobes, greenish to whitish gray.** .... **(p. 59)** *Parmelia bolliana*

202b **Upper surface without pores; thallus smaller with narrow, linear lobes.** 203

203a (202) **Margins of lobes with dense erect lobules (use hand lens and compare with Fig. 269); cortex K−.** .... **(p. 141)** *Phaeophyscia imbricata*

203b **Lobules absent or if present sparsely developed and adnate; cortex K+ yellow (except in species of *Lobaria*, *Phaeophyscia*, *Physciopsis*, and *Physconia*).** **204**

204a (203) **Collected on rocks or soil.** **205**

204b **Collected on trees.** **208**

205a (204) **Lobes very narrow, 0.3-0.6 mm wide. Fig. 210.** ...... *Physcia halei* Thoms.

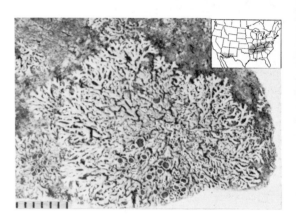

Figure 210

Figure 210   Physcia halei

Thallus whitish mineral gray, closely adnate, scattered, 2-3 cm broad; lower surface white, moderately rhizinate; apothecia common. Cortex and medulla K+ yellow (usually difficult to tell) (atranorin). Widespread on granite and sandstone in open areas. Some specimens of *Physcia subtilis* will key here if the soredia are sparse or overlooked. *Physcia phaea* (below) is much larger and has white spotting.

205b **Lobes 1-2 mm wide.** **206**

206a (205) Upper surface white-spotted (use hand lens and compare with Fig. 6E); cortex K+ yellow. Fig. 211. .... *Physcia phaea* (Tuck.) Thoms.

Figure 211

Figure 211   Physcia phaea (×2)

Thallus whitish mineral gray, adnate, 2-5 cm broad; lower surface white, moderately rhizinate; apothecia numerous. Cortex and medulla K+ yellow (atranorin and zeorin). Rather common on sheltered rocks in open woods. It is related to the corticolous *P. aipolia*. Another white-spotted species, *P. albinea* (Ach.) Nyl., grows on rocks at higher elevations in the western states; it has narrower lobes (no more than 1 mm wide) and reacts K− in the medulla. Another rock species in the West, *P. biziana* (see page 115), has a dull, finely white pruinose surface.

206b Upper surface without spotting, shiny or pruinose; cortex K−. 207

207a (206) Collected on soil or humus over rocks; thallus brownish gray, the surface shiny. .... (p. 147) *Phaeophyscia constipata*

207b Collected on rock; thallus whitish gray. .... (p. 141) *Anaptychia palmatula*

208a (204) Lobes 2-4 mm wide; thallus yellowish or greenish gray. 209

208b Lobes narrower, 0.5-2 mm wide; thallus whitish gray. 210

209a (208) Lower surface coarsely rhizinate; medulla often pale yellow, K+ yellow. .. .... (p. 48) *Pseudoparmelia sphaerospora*

209b Lower surface finely tomentose (use hand lens); medulla white, K−. .... (p. 132) *Lobaria ravenelii*

210a (208) Upper cortex scabrid and white pruinose (use hand lens and see Fig. 212), K−. 211

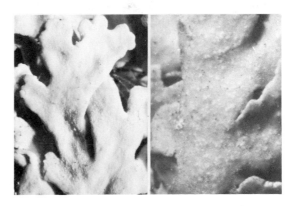

Figure 212

Figure 212   Pruina of *Heterodermia rugulosa* (left) (×10) and white spots of *Physcia aipolia* (about ×15).

210b Upper cortex smooth, usually shiny, not pruinose or with sparse pruina at lobe tips. 212

211a (210) Medulla uniformly white. ....
.... (p. 141) *Anaptychia palmatula*

211b Medulla (especially in apothecia) with orange-yellow patches (expose with razor blade). Fig. 213. ....
.. *Heterodermia rugulosa* (Kurok.) Hale

Figure 213

Figure 213   Heterodermia rugulosa

Thallus whitish gray, adnate on bark, 4-6 cm broad; lobes linear, 1-2 mm wide; upper surface at first smooth but soon becoming densely white pruinose; lower surface tan, moderately rhizinate; apothecia common, the rim lobulate-crenate. Cortex and medulla K+ yellow, the pigment K+ purple (atranorin and zeorin with anthraquinone pigments). Common at the base of oak trees in open woods and along roads from western Texas to Arizona. This is the only pigmented nonsorediate species of *Heterodermia*.

212a (210) Upper cortex white-spotted (use hand lens and see Fig. 212); medulla K+ yellow. Fig. 214. ....
.... *Physcia aipolia* (Ehrh.) Hampe

Figure 214

Figure 214   Physcia aipolia (×3)

Thallus whitish gray, closely adnate, 3-6 cm broad; lower surface white with moderate to dense to darkening rhizines, often projecting beyond the margin; apothecia very common, the disc black or becoming white pruinose. Cortex and medulla K+ yellow (atranorin and zeorin). Extremely common on exposed deciduous trees, often growing on planted trees in smaller cities. The thallus is thicker than in *Physcia stellaris* in addition to the differences in white-spotting and medullary K test. Separating these species at first, however, will prove troublesome. From North Carolina to Texas it may occur with *P. alba* (Fée) Müll. Arg., which has smaller, more appressed (lobes less than 1 mm wide).

212b Upper cortex continuous, without spots; medulla K+ or K−. 213

213a (212) Thallus closely appressed; lobes about 0.5 mm wide, confluent; cortex K−. Fig. 215. ....
.... *Physciopsis syncolla* (Tuck.) Poelt

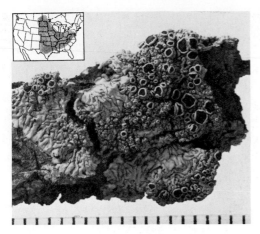

Figure 215

Figure 216

Figure 216 Heterodermia diademata

Figure 215 Physciopsis syncolla

Thallus 1-2 cm broad; lower surface sparsely rhizinate (difficult to determine in most specimens); apothecia common. Widespread in deciduous forests but most common on isolated trees in the prairie-forest states (Minnesota to Texas). It is easily overlooked because it blends with the bark. Rarely the medulla will be orange.

Thallus whitish gray, 3-6 cm broad; lobes 1-1.5 mm wide, crowded; lower surface light tan with sparse, long rhizines; apothecia numerous. Cortex and medulla K+ yellow (atranorin). Fairly common on oak trees in Arizona and New Mexico. This species, while externally identical with *H. hypoleuca*, has a lower cortex and rhizines. If an orange pigment is present, the collection can be identified as *H. rugulosa* (above), which has the same geographic range.

213b Thallus adnate; lobes 1-2 mm broad, separate; cortex K+ yellow. 214

214a (213) Rhizines long and branched (use lens); margins of apothecia dentate-lobulate. Fig. 216. ....
Heterodermia diademata (Tayl.) Awas.

214b Rhizines short and unbranched; rim of apothecia smooth, not dentate. 215

215a (214) Upper surface dull, finely white pruinose (use hand lens). Fig. 217. ....
.... *Physcia biziana* (Mass.) Zahlbr.

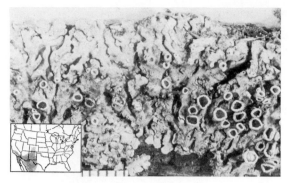

Figure 217

Figure 217 Physcia biziana

Thallus whitish gray, closely adnate on bark, 3-4 cm broad; lobes 1-2 mm wide, crowded; lower surface pale tan and rhizinate; apothecia common, the disc black or white pruinose. Cortex K+ yellow (atranorin); medulla K−. Widespread on oak, pine, and other trees, rarely on rock. While superficially resembling *P. stellaris,* this species has a dull uniformly pruinose cortex.

215b Upper cortex shiny, not pruinose. 216

216a (215) Medulla K−; apothecia blackish to white pruinose. Fig. 218. ....
.... *Physcia stellaris* (L.) Nyl.

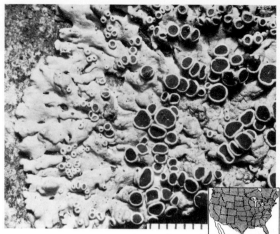

Figure 218

Figure 218 Physcia stellaris

Thallus closely adnate, 2-4 cm broad; below sparsely to moderately rhizinate; apothecia very common. Cortex K+ yellow (atranorin). Very common on deciduous trees, more rarely on conifers, in open woods or along roadsides. This species often grows in the canopy on small branches, while closely related *Physcia aipolia,* which has white spots, grows more toward the base. A KOH test is needed to tell these two apart.

216b Medulla K+ deep yellow; apothecia pale brown. Fig. 219. ....
...... *Parmeliopsis placorodia* (Ach.) Nyl.

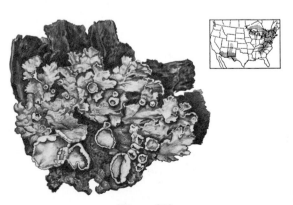

Figure 219

Figure 219   Parmeliopsis placorodia (×2)

Thallus whitish gray green, adnate, 2-4 cm broad; lower surface white, moderately rhizinate; apothecia numerous. Cortex and medulla K+ yellow, P+ yellow (thamnolic acid). Locally abundant in scrub pine forests and along roadsides. This species often grows with *Cetraria aurescens* and *Parmeliopsis aleurites* on dead branches.

BROWN LICHENS

217a (66) **Thallus soroidiate with marginal, apical, or laminal soralia (use hand lens and see Fig. 220); apothecia rarely found.** 218

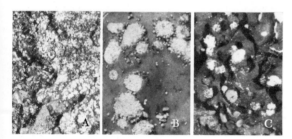

Figure 220

Figure 220   Soredia of *Parmelia subaurifera* (A), *Lobaria pulmonaria* (B), and *Parmelia sorediosa* (C) (×10)

217b Thallus isidiate or without soredia and isidia; apothecia rare to commonly found. 241

218a (217) **Thallus large with lobes 10-20 mm wide; lower surface mottled white and brown (do not use lens).** 219

218b Thallus small to medium-sized, the lobes 0.5-6 mm wide; lower surface uniformly white to tan, brown, or black, not mottled. 221

219a (218) **Upper surface strongly reticulately ridged (do not use lens). Fig. 221.** .... *Lobaria pulmonaria* (L.) Hoffm.

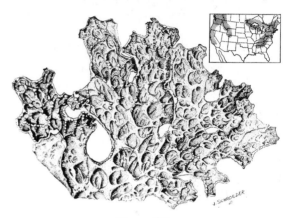

Figure 221

Figure 221   Lobaria pulmonaria (×1)

Thallus light brownish green, turning bright green when wet, loosely adnate, 5-25 cm broad; soredia often becoming coarsely isidiate (under lens); apothecia not common. Medulla K+ yellow, P+ pale orange (stictic and norstictic acids). Widespread in mature northern hardwood forests and swamps but becoming rare in reforested areas. Sparsely sorediate specimens in the West can be separated from *Lobaria linita* by the positive K test. This is the lichen which was widely used in the Middle Ages for treating lung diseases because of its resemblance to lung tissue.

219b Upper surface plane to weakly ridged or undulating. 220

220a (219) Upper surface weakly ridged to undulating; lower surface strongly mottled brown and white. Fig. 222. .... *Lobaria scrobiculata* (Scop.) DC.

Figure 222

Figure 222   Lobaria scrobiculata

Thallus light brownish green, loosely adnate, 6-12 cm broad; upper surface and margins sparsely sorediate; apothecia lacking. Medulla K+ yellow, KC+ red, P+ orange (stictic acid and scrobiculin). Rare at the base of trees and on mossy rocks in mature woods. Deforestation in the 19th century probably destroyed most of the habitats for this conspicuous lichen. In the northern Rocky Mountains one may be lucky enough to find *L. hallii* (Tuck.) Zahlbr., which is close except for having fine hairs and tomentum on the upper surface.

220b Upper surface plane; lower surface with a thick, uniform to weakly veined tomentum below. ....
.... (p. 50) *Peltigera collina*

221a (218) Lobes inflated, hollow (section with razor blade and see Fig. 125); rhizines lacking. ....
.... (p. 74) *Hypogymnia bitteri*

221b Lobes flat, solid, not appearing inflated; rhizines or tomentum sparse to well developed. 222

222a (221) Medulla conspicuously orange red (expose with razor blade). Fig. 223. ....
.... *Phaeophyscia rubropulchra* (Degel.) Moberg

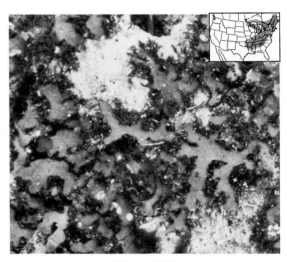

Figure 223

Figure   223   Phaeophyscia   rubropulchra (×10)

Thallus tannish green to light brown, closely adnate, colonies coalescing to form patches up to 10 cm broad or more; lobes short and crowded, 0.5-1 mm wide, black rhizines projecting along the margins; soralia irregular, laminal and terminal, the soredia coarse and sometimes isidioid; lower surface black and densely rhizinate; apothecia rare. Cortex K−; medulla K+ purple (skyrin). Very common on hardwood trees (*Ostrya*, elm, maple) in open forests and on rocks occasionally. This is the most easily recognized species of *Phaeophyscia* because of the red medulla, often exposed by slugs and snails which eat away the upper cortex.

222b Medulla white (pale yellow only in *Physconia enteroxantha*). 223

223a (222) Soredia diffuse over most of the upper surface and mixed with tiny isidia (use hand lens), leaving a white area when rubbed with fingers. Fig. 224. ......
.... *Parmelia subaurifera* Nyl.

Figure 224

Figure 224   Parmelia subaurifera

Thallus chestnut brown with a white cast, rather closely adnate on bark, 4-8 cm broad; lower surface light brown or blackening, moderately rhizinate; apothecia rare. Medulla K−, C+ red, P− (lecanoric acid). Very common on trees in open forests and along roads. This is a common brown *Parmelia* in eastern North America but also occurs less frequently throughout the western states. A very similar species, *P. glabratula* (Lamy) Nyl., has only

fine isidia without the soredia, and *P. subargentifera* (below), has distinct soralia and tiny cortical hairs on the lobe ends.

223b Soredia in delimited round or linear soralia; isidia lacking. 224

224a (223) Upper surface scabrid and becoming white pruinose (use hand lens and see Fig. 6D). Fig. 225. ....
.... *Physconia detersa* (Nyl.) Poelt

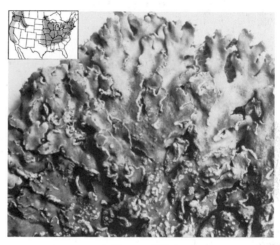

Figure 225

Figure 225   Physconia detersa (×3)

Thallus whitish gray, sometimes brownish, adnate on bark or over mosses, 4-7 cm broad; lobes 1-2 mm wide, marginally sorediate; lower surface black, densely rhizinate; apothecia rare. Cortex and medulla K− (no substances). Common on hardwood trees in open woods and along dusty roads, less common over mosses on rock (mostly in western states). This species had earlier been called "Physcia grisea." Closely related *P. enteroxantha* (Nyl.) Poelt is identical externally but has yellowish soredia and a pale yellow medulla. It has the same range as *P. detersa* but is much rarer.

224b Upper surface smooth, often shiny, lacking pruina. 225

225a (224) Lobes rather broad and rotund, 2-6 mm wide (use ruler). 226

225b Lobes narrower and linear, 0.3-2 mm wide. 229

226a (225) Lower surface bare, tan to brown. Fig. 226. .. *Nephroma parile* (Ach.) Ach.

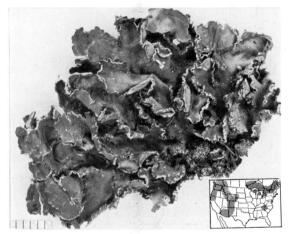

Figure 226

Figure 226   Nephroma parile

Thallus light brown or darker, loosely adnate among mosses, 4-10 cm broad; lower surface buff, bare; apothecia very rare. Cortex K−; medulla K−, C−, P− (unidentified substances). Widespread over mosses at the base of trees and on rocks in moist woods. One could mistake this lichen for *Parmelia subargentifera* (below), which has a black lower surface and rhizines.

226b Lower surface rhizinate or tomentose (use lens), brown or blackening. 227

227a (226) Upper surface with conspicuous white pores (use hand lens and compare with Fig. 85; actually visible with naked eye). Fig. 227. ....
.... *Parmelia stictica* (Del.) Nyl.

Figure 227

Figure 227   Parmelia stictica

Thallus light brown to brownish gray, closely adnate on rock, 4-6 cm broad; upper surface shiny, wrinkled with age; soredia developing from the white pores toward the center of the thallus; lower surface black at the center, dark brown at the margins, moderately rhizinate; apothecia lacking. Cortex K+ yellowish (atranorin); medulla K−, C+ rose, P− (gyrophoric acid). Rarely collected on rocks in exposed areas at higher elevation or along shorelines.

227b Upper surface uniform, without white pores. 228

228a (227) Lobes strap-shaped, suberect; shiny below with very sparse rhizines or mostly bare. Fig. 228. ....
.... *Cetraria chlorophylla* (Willd.) Vain.

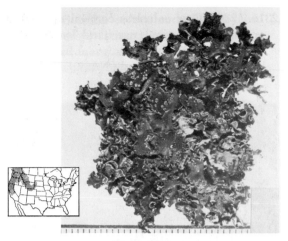

Figure 228

Figure 229

Figure 228 Cetraria chlorophylla

Thallus light brown to greenish brown or darker, 4-8 cm broad; soralia marginal **toward lobe tips**; lower surface lighter brown, shiny; apothecia lacking. Cortex K−; medulla K−, C−, P− (fatty acids). Common on conifers and fenceposts. This is one of the commonest foliose lichens in the western states. In Alberta, western North Dakota, and more rarely in the upper Great Lakes area, there is a very similar brown *Parmelia, P. albertana* Ahti; it can be separated by the more adnate, shorter lobes and marginal, labriform soralia. It reacts C+ red (lecanoric acid).

228b Lobes adnate, rotund; black below with moderate to dense rhizines. Fig. 229. .... *Parmelia subargentifera* Nyl.

Figure 229 Parmelia subargentifera

Thallus chestnut to olive brown, adnate, 6-12 cm broad; soralia marginal and in part laminal; apothecia rare. Medulla C+, KC+ red (lecanoric acid). Rather common on trees in open woods and swamps, less so on rocks. It is rarer than *P. subaurifera* but can be recognized by the discrete soralia and tiny cortical hairs at lobe tips. *Parmelia albertana*, which will also key here, is smaller with labriform soralia. *Parmelia stictica* may key out here, too; it has white pores in the upper cortex (see page 120).

229a (225) Collected on trees. 230

229b Collected on rocks. 235

230a (229) Lower surface light tan to white (use hand lens). Fig. 230. .... *Physcia chloantha* Ach.

234a (233) Lobes about 0.5 mm wide (use ruler); soralia mostly terminal on small lobes. Fig. 234. .... *Phaeophyscia pusilloides* (Zahlbr.) Essl.

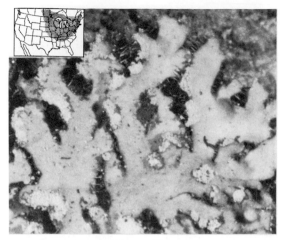

Figure 234

Figure 234   Phaeophyscia pusilloides (×10)

Thallus greenish to brownish gray, closely adnate on bark, 1-3 cm broad; lobes narrow, 0.5-1 mm wide, separate, with white-tipped rhizines projecting along the margins; soralia small, orbicular, mostly subterminal; lower surface black and densely rhizinate; apothecia rare. Cortex and medulla K−, C−, P− (no substances.) Very common on hardwood trees in open woods. Another member of the "Physcia orbicularis" complex, this species is characterized by the small capitate soralia. Forms of *P. hispidula* with capitate soralia resemble it but are much larger with lobes 1-2 mm wide.

234b Lobes 1.5-2 mm wide; soredia laminal. Fig. 235. ....
Phaeophyscia hispidula (Ach.) Moberg

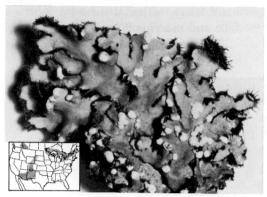

Figure 235

Figure 235   Phaeophyscia hispidula (×3)

Thallus greenish mineral gray, adnate, 2-4 cm broad; lower surface black, densely rhizinate, the rhizines projecting out along the margins; apothecia lacking. Cortex and medulla K−, C−, P− (no substances). Rather rare on trees in open forests in eastern United States but common in the western states. This species is not likely to be confused with any other Phaeophyscias because of the larger lobes and capitate laminal soralia.

235a (229) Lower surface tan to white. *Massalongia carnosa* (see below under *Phaeophyscia sciastra* and (p. 122) *Physcia chloantha*.

235b Lower surface brown or blackening. ......
.... 236

236a (235) Soredia marginal, isidioid, dark; thallus dark to blackish mineral gray. Fig. 236. ....
.. *Phaeophyscia sciastra* (Ach.) Moberg

Figure 236

Figure 236 Phaeophyscia sciastra

Thallus closely adnate, 2-5 cm broad; lobes narrow, less than 1 mm wide; upper surface sometimes becoming white pruinose; lower surface black, moderately rhizinate; apothecia rare. Cortex and medulla, K−, C−, P− (no substances). Widespread on exposed rocks, especially near lakes and streams. The thallus color often blends with the rock substratum. It is also often difficult to collect from the rock surface. A superficially similar but more distinctly brown lichen with blue-green algae, *Massalongia carnosa* (Dicks.) Arn., will be collected on mosses over rocks at higher elevations in the western states.

236b Soredia powdery, white (blackening in *Parmelia disjuncta* and *P. sorediosa*), laminal marginal, or terminal; thallus brown to black or greenish gray. 237

237a (236) Thallus greenish gray. ....
.... (p. 123) *Phaeophyscia adiastola*

237b Thallus chestnut brown to blackish. ......
.... 238

238a (237) Thallus dull (use hand lens); soralia mostly terminal on short lobes. Fig. 237. ....
.... *Parmelia sorediosa* Almb.

Figure 237

Figure 237 Parmelia sorediosa (×2)

Thallus dark brown, closely adnate on rock and usually collected with the rock, 2-4 cm broad; lobes about 0.5 mm wide; upper surface weakly foveolate; lower surface dark brown, rhizinate; apothecia lacking. Medulla K−, C−, P− (perlatolic acid). Common on acidic rocks in ledges and stonewalls. *P. disjuncta* is often mistaken for this species but it has mostly laminal soralia and is shiny.

238b Thallus shiny; soralia laminal or marginal. 239

239a (238) Medulla C+ rose; pseudocyphellae conspicuous on upper surface. Fig. 238. ....
.... *Parmelia substygia* Räs.

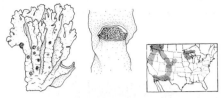

Figure 238

Figure 238 Parmelia substygia (×4)

Thallus dark brown, closely adnate, 3-8 cm broad; upper surface shiny; lower surface sparsely rhizinate, black; apothecia rare. Medulla C+, KC+ red (gyrophoric acid). Widespread on boulders, rarely on wood, in open areas. This species is identified by the C+ test and by the small but conspicuous white pseudocyphellae on the lobe surface. It is often confused with *P. disjuncta* when the C test is not done properly.

239b Medulla C−; pseudocyphellae inconspicuous or lacking. 240

240a (239) Soralia laminal; medulla K−. Fig. 239. .... *Parmelia disjuncta* Erichs.

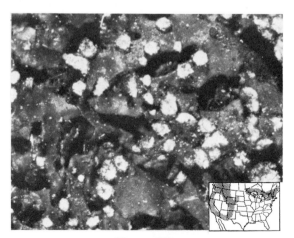

Figure 239

Figure 239 Parmelia disjuncta (×10)

Thallus dark brown, closely adnate on rock, 2-4 cm broad; lobes 0.5-1 mm wide, crowded; soralia mostly laminal toward center of thallus; lower surface dark brown to black, rhizinate; apothecia lacking. Medulla K−, C−, P− (perlatolic acid). Common on rocks, ledges, and stonewalls in open woods. In general, *P. disjuncta* is more common in the western states than *P. sorediosa*, the most closely related species. If the soredia originate from large hollow pustules you may have *Neofuscelia loxodes* (see page 128).

240b Soralia marginal; medulla K+ yellow turning red. Fig. 240. .... *Cetraria culbersonii* Hale

Figure 240

Figure 240 Cetraria culbersonii

Thallus chestnut brown, adnate on rock, 3-6 cm broad; lobes linear, 1-2 mm wide; lower surface brown, moderately rhizinate; apothecia lacking. Medulla K+ yellow turning red, P+ orange (stictic and norstictic acids). Widespread on rocks in open areas and in talus slopes in the Appalachian Mountains. This distinctive species is the sorediate form of *Cetraria hepatizon.*

241a (217) Isidia present, cylindrical, flattened, or warty (use hand lens and see Fig. 241) (do not mistake erect marginal pycnidia of *Cetraria* species for isidia); apothecia rare. 242

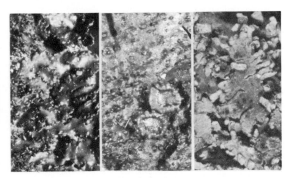

Figure 241

Figure 241 Isidia of *Cetraria coralligera* (A), papillate warts of *Parmelia exasperata* (B) and (C), isidia of *Pannaria tavaresii* (×10)

241b Isidia lacking, the thallus surface smooth to wrinkled or warty (marginal lobules present in *Anaptychia palmatula*, *Nephroma helveticum*, *Phaeophyscia imbricata*); apothecia commonly developed. 252

242a (241) Isidia thin and cylindrical (use hand lens and see Fig. 241) (becoming hollow and pustulate in *Neofuscelia loxodes* only). 243

242b Isidia flattened, lobulate, or thick, warty or granular. 248

243a (242) Lower surface tomentose (use hand lens); isidia mostly marginal, white-tipped. Fig. 242. .... *Pannaria tavaresii* Jørg.

Figure 242

Figure 242 Pannaria tavaresii

Thallus brownish mineral gray to light brown, closely adnate, 4-6 cm broad; isidia mostly marginal, whitish pruinose at the tips; lower surface light brown, densely tomentose; apothecia rare. Medulla P+ red (pannarin) or P−. At the base of trees or on mosses over rocks in mature woods. When very closely adnate it resembles *Pannaria leucosticta*, which lacks distinct lobation. The algae are blue-green, and this character, along with tomentum, will separate it from species of *Parmelia* or *Anaptychia.*

243b Lower surface sparsely to moderately rhizinate, the lower cortex easily seen between rhizines; isidia mostly laminal, brown. 244

244a (243) Lobes very narrow, 0.5-0.8 mm wide (use ruler); thallus 1-3 cm broad. Fig. 243. .... *Cetraria coralligera* (Weber) Hale

Figure 243

Figure 243   Cetraria coralligera

Thallus dark brown, closely adnate; isidia dense, cylindrical; lower surface light brown, sparsely rhizinate; apothecia lacking. Medulla K−, C−, P− (fatty acids). Widespread on dead trees and fenceposts. *Cetraria fendleri* is somewhat similar but lacks isidia. The isidiate brown Parmelias are much larger in size.

244b Lobes broader, 1-6 mm wide; thallus 3-8 cm broad. 245

245a (244) Upper surface weakly reticulately ridged and with white markings (use hand lens toward lobe tips and see Fig. 137); thallus mineral gray but turning brown at lobe tips; medulla K+ yellow turning red. .... (p. 89) *Parmelia saxatilis*

245b Upper surface plane, without white markings; thallus uniformly brown; medulla K−. 246

246a (245) Isidia hollow, inflated, breaking open apically and appearing sorediate (use hand lens and see Fig. 241). Fig. 244. .... *Neofuscelia loxodes* (Nyl.) Essl.

Figure 244

Figure 244   Neofuscelia loxodes

Thallus chestnut brown, closely adnate on rocks, 2-4 cm broad; lobes crowded, about 1 mm wide, the surface wrinkled, finely reticulately marked at the tips (use lens), the wrinkles developing into coarse, thick isidia that burst open; lower surface dark brown to black; apothecia lacking. Medulla K−, C−, P− (perlatolic, glomelliferic, and glomellic acids). Widespread, but not commonly collected, on rocks in arid regions. The lack of pseudocyphellae and the pustules separate this lichen from the more common *Parmelia disjuncta*. *Neofuscelia subhosseana* (Essl.) Essl. (K+ red, unknown substances) is a virtually identical species known only from California to Washington, as is also *N. verruculifera* (Nyl.) Essl., which has divaricatic acid (medulla K−, C−, P−) and occurs in the western states.

246b Isidia thin and solid, not breaking open (see Fig. 241). 247

247a (246) Medulla C+ red; isidia very fine. .. .. *Parmelia glabratula* (see *P. subaurifera*, page 119)

247b Medulla C−; isidia normal. Fig. 245. .... *Parmelia elegantula* (Zahlbr.) Szat.

Figure 245

Figure 245   Parmelia elegantula

Thallus chestnut brown to olive green, adnate, 3-8 cm broad; upper surface often becoming white pruinose; isidia cylindrical, often branched; lower surface brown or blackening; apothecia rare. Medulla K−, C−, P− (no substances). On trees and sheltered rocks. When occurring on rocks it must be carefully distinguished from the related *P. infumata* Nyl. which has isidia without apical pseudocyphellate initials. Another corticolous species occurring from California to Alberta, *P. subelegantula* Essl. also has this same type of isidia but some become lobulate with age. This species would also intergrade with *Parmelia*

exasperatula (below), which has mostly flattened isidia. Rare *Neofuscelia chiricahuensis* (Anders. & Weber) Essl., a saxicolous species from Arizona, is superficially close but has simple to branched isidia and a K+ red reaction (stictic and norstictic acids).

248a (241) Isidia flattened and lobulate (use hand lens). 249

248b Isidia papillate or granular (see Fig. 241). 251

249a (248) Isidia mostly laminal, thallus dark greenish chestnut brown. Fig. 246. .... *Parmelia exasperatula* Nyl.

Figure 246

Figure 246   Parmelia exasperatula (×10)

Thallus adnate, 3-6 cm broad; upper surface shiny, plane; lower surface black, moderately rhizinate; apothecia rare. Medulla K−, C−, P− (no substances). Widespread on trees in open woods and on sheltered rocks. *Parmelia elegantula* is close but has cylindrical isidia. *Parmelia panniformis* (Nyl.) Vain., a widespread but not commonly collected saxicolous species in northern areas, has very dense, flattened isidia that may hide the main thallus lobes. It contains perlatolic acid (medulla K−, C−, P−) that would have to be confirmed with TLC. Some forms of *P. subelegantula* with lobulate isidia will key here (see under *P. elegantula*) but most of the isidia are cylindrical.

249b Isidia (lobules) marginal; thallus light brown to chestnut brown. 250

250a (249) Lower surface rhizinate; lobes 1-2 mm wide. .... **(p. 141)** *Phaeophyscia imbricata*

250b Lower surface bare; lobes 2-6 mm wide. .... **(p. 134)** *Nephroma helveticum*

251a (248) Thallus suberect; collected in western North America. .... **(p. 133)** *Cetraria platyphylla*

251b Thallus closely adnate; collected in eastern North America. .... **(p. 136)** *Parmelia exasperata*

252a (241) Lower surface and medulla brilliant orange red (do not use hand lens). Fig. 247. .... *Solorina crocea* (L.) Ach.

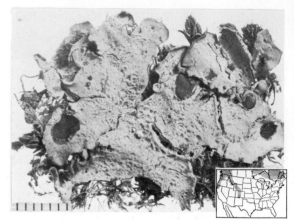

Figure 247

Figure 247   Solorina crocea

Thallus light greenish brown, adnate, 4-6 cm broad; medulla orange red; lower surface lacking a cortex but with distinct veins; apothecia common, located in sunken pits on the surface. Medulla K+ purple (solorinic acid). Common on soil. This unusual arctic lichen cannot be mistaken for any other species.

252b Lower surface white, tan, brown, or black; medulla white. 253

253a (252) Lobes very broad, 10-25 mm wide, the surface broadly reticulately ridged (without lens); lower surface mottled brown-white. Fig. 248. .... *Lobaria linita* (Ach.) Rabh.

Figure 248

Figure 248   Lobaria linita

Thallus light greenish to brownish gray, loosely attached to suberect, 6-20 cm broad; apothecia common. Cortex K−; medulla K−, C−, P− (tenuiorin). Common in moist fir forests of the Pacific Northwest. *Lobaria pulmonaria* is the same size but has soredia. The two species will occur together.

253b **Lobes narrower, 0.5-10 mm wide, the surface plane to wrinkled; lower surface uniformly colored, not mottled.** 254

254a (253) **Lower surface tomentose, pale brown (use hand lens and compare with Fig. 7).** 255

254b **Lower surface bare and shiny or rhizinate, white, brown, or black with cortex easily visible between rhizines (tomentum if present bluish black; only *Anaptychia setifera* is ecorticate).** 257

255a (254) **Apothecia located on the lower surface of lobe tips; tomentum intermixed with large white papillae (lens not needed). Frontispiece No. 12 and Fig. 249.** ....
.... *Nephroma resupinatum* (L.) Ach.

Figure 249

Figure 249   Nephroma resupinatum

Thallus light greenish brown, loosely adnate, 4-8 cm broad; lobes to 8 mm broad, ascending at the tips when apothecia present, sometimes short lobulate along margins with age; apothecia to 8 mm in diameter. Cortex and medulla K− (unidentified substances). Widespread on trees and rocks in moist woods.

255b **Apothecia (if present) adnate on the upper surface of lobes; papillae absent below.** 256

256a (255) **Lobes 3-6 mm wide, fragile; upper surface rugose and wrinkled (lens not needed). Fig. 250.** ....
.... *Lobaria ravenelii* (Tuck.) Yoshim.

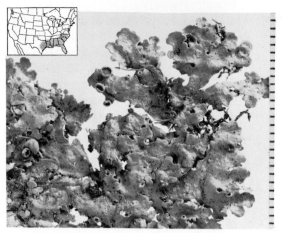

Figure 250

Figure 250   Lobaria ravenelii

Thallus light brown, loosely adnate on bark, turning green when wet, 5-10 cm broad; lobe margins becoming coarsely lobulate, crowded; lower surface tan, tomentose but with bare areas; apothecia common, the rim crenate-lobulate. Cortex K—; medulla K—, C+ rose, P— (gyrophoric acid). Common on trees (oak, beech, magnolia) in open woods, rarely on mosses over rocks, in the Coastal Plain. This is a smaller, darker lichen than *L. quercizans,* which occurs farther north in the Piedmont and mountains. Another subtropical *Lobaria, L. tenuis* Vain., occurs in Florida and Georgia; it has numerous lobules toward the center of the thallus.

256b Lobes 5-10 mm wide, firm; upper surface rugose only toward the center. Fig. 251. .... *Lobaria quercizans* Michx.

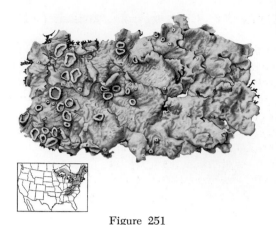

Figure 251

Figure 251   Lobaria quercizans (×1)

Thallus light brownish mineral gray, turning bright green when wet, 6-20 cm broad; lower surface tan, felty, with some tufts of tomentum; apothecia common. Medulla C+, KC+ rose (gyrophoric acid). Common on deciduous trees, especially maples, in open woods and swamps, and on rocks. This conspicuous lichen will at first be confused with a *Parmotrema* or *Pseudoparmelia* species, but the lower surface is tan and felty rather than rhizinate.

257a (254) Collected on wet rocks (or near the water line) in rivers and lakes. Fig. 252. .... *Dermatocarpon fluviatile* (G. Web.) Th. Fr.

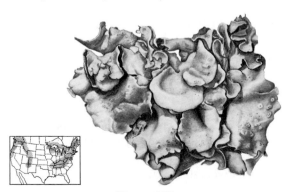

Figure 252

Figure 252   Dermatocarpon fluviatile (×1)

Thallus light brown, adnate to loosely attached, 1-4 cm broad; lobes irregular, often crowded; upper surface with black dots (perithecia); lower surface bare, wrinkled, blackish at the center but brown toward the margin. Widespread on moist or wet rocks throughout North America. This lichen may form a distinct zone or band on rocks just above water level. It intergrades with *Dermatocarpon miniatum,* which is usually distinctly umbilicate, grows on drier rocks, and does not turn green when wet.

257b Collected on trees or rocks in dry habitats. 258

258a (257) Collected on trees or on mosses at the base of trees. 259

258b Collected on rocks, over mosses on rocks, or on soil. 280

259a (258) Lobes rather broad and rotund, 2-6 mm wide (use ruler) (may be up to 10 mm in *Nephroma* species); thallus adnate to loosely attached or suberect; lower surface bare or rhizinate (use hand lens). 260

259b Lobes narrow and linear, 0.3-1.5 mm wide; thallus always adnate to appressed; lower surface rhizinate or tomentose, rarely bare. 269

260a (259) Margins of lobes (and sometimes surface) with erect, conspicuous pycnidia (use hand lens and see Fig. 5C); thallus loosely adnate to suberect. 261

260b Margins of lobes smooth; pycnidia (if present) immersed, laminal; thallus adnate. 262

261a (260) Thallus wrinkled and papillate; lobes to 10 mm wide, without cilia. Fig. 253. .... *Cetraria platyphylla* Tuck.

Figure 253

Figure 253   Cetraria platyphylla

Thallus chestnut brown, 3-8 cm broad; upper surface and rim of apothecia rugose, warty or subpapillate to coarsely isidiate; lobe margins becoming dissected; lower surface wrinkled, light brown, sparsely rhizinate; apothecia common. Medulla K−, C−, P− (fatty acids). Very common on conifers in open forests. The wrinkled surface and deep brown color separate this species from *Cetraria ciliaris,* and *C. merrillii* is much smaller and olive black. It is one of the commonest lichens in the northern Rocky Mountains and Cascades.

261b Thallus shiny and smooth, without papillae; lobes 1-4 mm wide, sparsely ciliate. Frontispiece No. 11 and Fig. 254. ....
.... *Cetraria ciliaris* Ach.

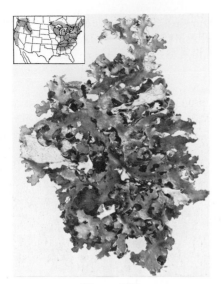

Figure 254

Figure 254   Cetraria ciliaris (×1.5)

Thallus light greenish to olive brown or darker, 3-7 cm broad; upper surface often weakly wrinkled (with lens); lower surface white to buff; apothecia numerous. Cortex K−; medulla K−, C+ red, P− (olivetoric acid). Very common on conifers, hardwoods, and fenceposts in open woods or along roads. This species is especially common in the Appalachian Mountains. In the Great Lakes area a virtually identical species with different chemistry (medulla C−, UV+ white, alectoronic acid), *C. halei* Culb., is more common. A somewhat smaller but similar species, *C. orbata* (Tuck.) Nyl., is most common in the western states, differentiated by fatty acids (medulla UV−, K−, C−, P−) and lack of cilia (small spinules often present). In the western states, however, the group is overshadowed by *C. platyphylla*.

262a (260) Lower surface bare (use hand lens). 263

262b Lower surface with rhizines. 264

263a (262) Margins of lobes lobulate to dentate-isidiate. Fig. 255. .... *Nephroma helveticum* (Ach.) Ach.

Figure 255

Figure 255   Nephroma helveticum

Thallus uniformly light brown, loosely adnate, 4-8 cm broad; lower surface light brown or darkening; apothecia common, located on the lower side of lobe tips. Widespread at the base of trees and on mossy rocks in mature woods. This is the most commonly collected *Nephroma* in the United States. Though completely unrelated, *Anaptychia palmatula* may key out here. It has much narrower lobes 1-2 mm wide, rhizines below, and laminal apothecia.

263b Margins of lobes entire. Fig. 256. .... *Nephroma bellum* (Spreng.) Tuck.

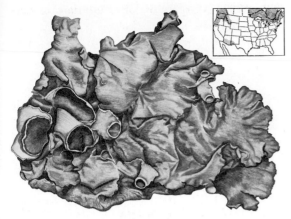

Figure 256

Figure 256   Nephroma bellum (×1.5)

Thallus light brown, loosely attached to adnate, 4-8 cm broad; lower surface tan or darkening, smooth and bare; apothecia common. Medulla K— or K+ faint yellow (unknowns). At the base of trees or on rocks in open woods or along roadsides. A very close relative, *N. laevigatum* Ach., has a similar range but the medulla is distinctly pale yellow and K+ deep yellow.

264a (262) Thallus loosely adnate; lobes sparsely ciliate. ....
.... (p. 134) *Cetraria ciliaris*

264b Thallus closely adnate; lobes lacking cilia. 265

265a (264) Collected in western North America. 266

265b Collected in eastern North America. ......
.... 267

266a (265) Medulla C+ red; minute hairs on lobe tips and rim of the apothecia (use hand lens). Fig. 257. ....
.... *Parmelia glabra* (Schaer.) Nyl.

Figure 257

Figure 257   Parmelia glabra

Thallus chestnut brown, adnate on bark, 4-7 cm broad; lobes rather broad, 2-4 mm wide, sometimes hidden by apothecia; lower surface dark brown or blackening; apothecia numerous. Medulla K—, C+, P— (lecanoric acid). Common on oaks and alders in California. A closely related species, rarer *P. glabroides* Essl., has the same chemistry but lacks the hairs and always grows on rocks. It has been collected in California, Colorado, and Washington.

266b Medulla C—; hairs lacking. Fig. 258. ......
.... *Parmelia subolivacea* Nyl.

Figure 258

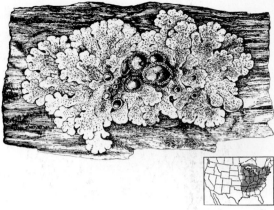

Figure 259

Figure 259 Parmelia exasperata (×2)

Thallus dark brown, closely adnate, 2-4 cm broad; lower surface dark brown to black, moderately rhizinate; apothecia common. Medulla K−, C−, P− (no substances). Widespread, but never abundant, on branches and trunks of trees in open woods. This lichen is most frequently collected in the southeastern states.

267b Lobe surface and apothecial rim smooth to rugose. 268

268a (267) Lower surface black; small white pores (pseudocyphellae) numerous on the supper surface (use hand lens). Fig. 260. .... *Parmelia olivacea* **(L.) Ach.**

Figure 258 Parmelia subolivacea

Thallus brown, closely adnate, 3-6 cm broad; lower surface brown or darkening, moderately rhizinate; apothecia common. Medulla K−, C−, P− (no substances). Widespread on conifers and other trees in open woods and along roads. This is brown the commonest *Parmelia* in the western states. Another virtually identical species, *P. multispora* Schneid., has asci with 16 rather than 8 spores in each ascus, a character that must be determined with a microscope. All fertile specimens that react K−, C−, P− should be examined for spores. Specimens on rocks with this chemistry are probably *P. glabroides*, mentioned above under *P. glabra*.

267a (265) Lobe surface and rim of apothecia with warty papillae. Fig. 259. .... *Parmelia exasperata* **DeNot.**

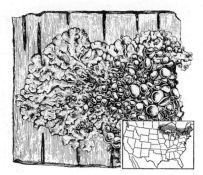

Figure 260

Figure 260   Parmelia olivacea (×1)

Thallus brown to olive greenish brown, adnate, 3-7 cm broad; lower surface black, sparsely rhizinate; apothecia very common, the rim thick. Medulla K−, C−, P+ red (fumarprotocetraric acid). Common on deciduous trees and conifers in open woods, especially in the Great Lakes region. The old *Parmelia olivacea* group has been split into several species. Almost all specimens in the central and southern Appalachians are *P. halei*, a smaller species. Two species with a pale lower surface are discussed below: *P. septentrionalis* and *P. trabeculata*.

268b Lower surface tan to brown; white pores lacking or very sparse. Fig. 261. ....
.. Parmelia septentrionalis (Lynge) Ahti

Figure 261

Figure 261   Parmelia septentrionalis

Thallus brown to olive or greenish brown, closely adnate, 2-6 cm broad; upper surface with some white pores (high power lens), shiny; lower surface light brown, rarely darkening, moderately rhizinate; apothecia numerous, the rim thin. Medulla P+ red (fumarprotocetraric acid). Common on trees in open woods, especially on alder, birch, and aspen. This species is especially common in the Great Lakes area and will usually be found in herbarium collections under the name "Parmelia olivacea." A rarer species in the boreal forests in Canada is close: *P. trabeculata* Ahti which reacts K+ red (norstictic acid) and has a tan, reticulate lower surface.

269a (259) Upper surface scabrid and white pruinose (use hand lens and see Fig. 6D). Fig. 262. ....
Physconia pulverulenta (Schreb.) Poelt

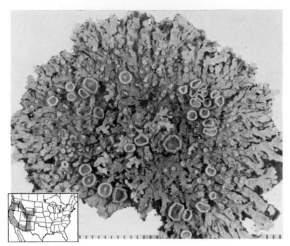

Figure 262

Figure 262 Physconia pulverulenta

Thallus brownish to whitish gray, adnate, 3-6 cm broad; lower surface black, moderately rhizinate; apothecia common. Cortex K—; medulla K—, C—, P— (no substances). Widespread on oaks and other trees in open areas. All collections from eastern North America previously identified as this species are apparently *Anaptychia palmatula* (see page 141), which has a white or tan lower surface. The K— test on the cortex will separate it from large forms of *Physcia aipolia*.

269b Upper surface smooth, usually shiny, not pruinose. 270

270a (269) Lower surface bare or rhizinate (lower cortex visible between rhizines) (use hand lens). 271

270b Lower surface densely bluish black tomentose (use hand lens and see Fig. 7). 278

271a (270) Margins of lobes and/or apothetical rim with adnate to erect black pycnidia (use hand lens and see Fig. 5C); lower surface whitish to tan. 272

271b Margins of lobes lacking pycnidia; lower surface generally dark brown or black (white to tan only in *Anaptychia setifera, A. palmatula, Phaeophyscia constipata,* and *Solorina saccata*). 275

272a (271) Thallus loosely adnate, 2-6 cm broad; sparse cilia present on lobe margins. .... (p. 134) *Cetraria ciliaris*

272b Thallus adnate (suberect only in *Cetraria merrillii*), only 1-2 cm broad. .. 273

273a (272) Lobes becoming suberect; thallus olive black. Fig. 263. .... *Cetraria merrillii* DR.

Figure 263

Figure 263 Cetraria merrillii

Thallus brownish to olivaceous black, adnate to somewhat tufted and suberect on twigs, 1-2 mm broad; lobes about 1 mm wide, becoming dissected; lower surface wrinkled, paler brown, sparsely rhizinate; apothecia common, the rim crenate-dentate. Cortex and medulla K−, C−, P− (fatty acids). Widespread on branches of conifers in more exposed areas. This typically western species is much smaller and darker than *C. ciliaris* and *C. orbata*. *Cornicularia californica* (Tuck.) DR., a poorly known western lichen, is superficially very close but lacks any marginal pycnidia, and has a prosoplectenchymatous lower cortex.

273b Lobes closely adnate; thallus chestnut brown. **274**

274a (273) Lobes finely dissected at the tips; apothecia not numerous; thallus turning green when wet. Fig. 264. .... *Cetraria fendleri* (Nyl.) Tuck.

Figure 264

Figure 264   Cetraria fendleri

Thallus brown, very closely adnate; 1-2 cm broad; lobes about 0.5 mm wide; lower surface tan, sparsely rhizinate; apothecia common. Cortex K−; medulla K−, C−, P− (fatty acids). Common on branches and trunks of conifers in open woods, often overlooked because it blends with color of the bark. Juvenile

specimens of *C. ciliaris* will have lobes at least 1 mm wide and not dissected. A close relative in New Mexico and Arizona, which was previously identified with *C. fendleri*, reacts C+ red (olivetoric acid) and has slightly broader, more linear lobes without dissected tips; it is now called *C. weberi* Essl.

274b Lobes entire, 1-2 mm wide, often hidden by well developed apothecia; thallus not changing color when wet. Fig. 265. .... *Cetraria sepincola* (Ehrh.) Ach.

Figure 265

Figure 265   Cetraria sepincola (×1.5)

Thallus dark brown, adnate, sometimes easily plucked from twigs, round, about 1 cm broad; margins of lobes rarely ciliate; lower surface tan, sparsely rhizinate; apothecia numerous. Cortex and medulla, K−, C−, P− (fatty acids). Common on twigs of alder, birch, and other deciduous trees and conifers, often in bogs. Though much smaller than *C. orbata*, these two species seem to intergrade and a few specimens will be difficult to identify. *Cetraria merrillii* has suberect lobes with less conspicuous apothecia.

275a (271) Thallus chestnut brown, not changing color significantly when wetted. Fig. 266. .... *Parmelia halei* Athi

Figure 266

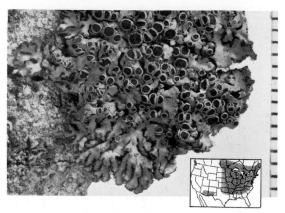

Figure 267

Figure 266 Parmelia halei

Thallus closely adnate on bark, 3-6 cm broad; lobes rather narrow and becoming dissected or lobulate, about 1 mm wide; upper surface shiny, becoming wrinkled with age; lower surface dark brown to black, sparsely rhizinate; apothecia common, small. Medulla K−, C−, P+ red (fumarprotocetraric acid). Common on hardwoods, especially *Acer*, in open woods. This is the most frequently collected non-isidiate, nonsorediate brown *Parmelia* in the Appalachian Mountains.

275b Thallus pale brownish gray, turning green when wetted. **276**

276a (275) Margins of lobes and apothecial rim entire, without lobules. Fig. 267. ...... *Phaeophyscia ciliata* (Hoffm.) Moberg

Figure 267 Phaeophyscia ciliata

Thallus brownish to greenish gray, closely adnate, 2-3 cm broad; lobes about 1 mm wide; lower surface black, sparsely to densely rhizinate; apothecia common. Cortex and medulla K−, C−, P− (no substances). Widespread on deciduous trees, especially aspens, in open woods and along roads. A less common variant with a red medulla (skyrin) may be identified as *P. erythrocardia* (Tuck.) Essl. Robust specimens may resemble *Physcia stellaris*, which differs in having a white lower surface and a K+ yellow cortex. *Physciopsis syncolla* (see page 115) is much more closely appressed and also pale below without rhizines. On careful examination, some specimens will be found to have fine colorless hairs on the lobe margins and apothecia; these can be identified as *Phaeophyscia hirtella* Essl.

276b Margins of lobes (and apothecia if fertile) lobulate (use hand lens). **277**

277a (276) Lobes appressed; apothecia common. Fig. 268. ....
.. *Anaptychia palmatula* (Michx.) Vain.

Figure 268

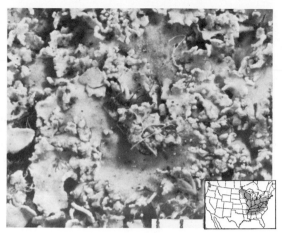

Figure 269

Figure 269 Phaeophyscia imbricata

Figure 268 Anaptychia palmatula (×3)

Thallus light brown, rarely white pruinose, turning deep green when wet, adnate, 4-8 cm broad; lobes linear, crowded, 1-2 mm wide; lower surface whitish tan, moderately rhizinate; apothecia common, the rim lobulate. Cortex and medulla K−, C−, P− (no substances). Common at the base of deciduous trees and on shaded rocks in mature forests. The white pruinose sun-form of this species has often been misidentified as *Physconia pulverulenta* or *P. leucoleiptes* in the eastern states.

277b Lobules erect; apothecia rarely found. Fig. 269. .... **.... *Phaeophyscia imbricata* (Vain.) Essl.**

Thallus greenish to brownish gray, adnate, 4-8 cm broad; upper surface shiny or becoming white pruinose in part; lower surface dark at the center, whitish toward the tips, densely rhizinate; apothecia rare. Cortex and medulla K−, C−, P− (no substances). Common at the base of deciduous trees and over mosses on trees in mature forests. If the lobules are poorly developed it might be mistaken for *Phaeophyscia ciliata*, which has a black lower surface and is more appressed. *Parmeliella pannosa*, which may also be quite lobulate, has dense black tomentum below.

278a (270) Lobes 1-3 mm wide (use ruler); thallus loosely adnate. Fig. 270. .... *Pannaria lurida* (Mont.) Nyl.

Figure 270

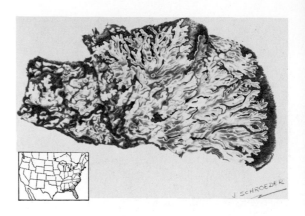

Figure 271

Figure 270   Pannaria lurida (×1.5)

Thallus dull brownish, 4-6 cm broad; lower surface buff, with long conspicuous tomentum; algae blue-green; apothecia common. Medulla P+ orange red (pannarin) or rarely P−. On oaks and junipers in exposed woods, near cliffs, or at the base of trees along roadsides. This species is not often collected but is easily recognized by the tomentum. *Pannaria rubiginosa* (below) intergrades with it but is usually smaller and more adnate.

278b Lobes 0.5-1 mm wide; thallus closely adnate. 279

279a (278) Thick tomentum projecting as a mat around lobes; apothecia lacking. Fig. 271. ....
.... *Parmeliella pannosa* (Sw.) Nyl.

Figure 271   Parmeliella pannosa

Thallus closely adnate, brownish gray, 4-8 cm broad; upper surface becoming minutely lobulate along the margins; algae blue-green; apothecia usually lacking. Cortex and medulla K−, C−, P− (no substances). Rather rare on deciduous trees in hammocks and pastures. This is a tropical species that may be disappearing as forests are cut down. Another species in this genus, *P. plumbea* (Lightf.) Müll. Arg., has a much larger thallus; it is very rarely collected on trees and rocks from Maine to Nova Scotia.

279b Tomentum not projecting conspicuously at the margin; apothecia common, the rim crenate. Fig. 272. ....
.... *Pannaria rubiginosa* (Ach.) Del.

Figure 272

Figure 272 Pannaria rubiginosa

Thallus light brownish gray, closely adnate, 2-5 cm broad; upper surface often white pruinose at the margins, becoming short lobulate; lower surface variable, dark brown or lighter; algae blue green; apothecia common. Cortex and medulla K−, C−, P− (no substances). Widespread at the base of trees and on rocks in mature forests. *Pannaria lurida* has much broader lobes and reacts P+ red. *Pannaria leucostica* (see page 233) is a squamulose lichen in this genus.

280a (258) **Margins of lobes with long cilia (hand lens not needed). Fig. 273.** ....
.... *Anaptychia setifera* Räs.

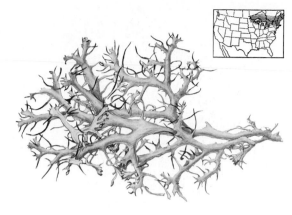

Figure 273 Anaptychia setifera (×2)

Thallus loosely adnate to almost subfruticose, 4-8 cm broad; lower surface white, lacking a cortex; apothecia not common. Cortex and medulla K−, C−, P− (no substances). Common on sheltered rocks or among shrubs, especially along lake shores. The only comparable lichen, *Phaeophyscia constipata* (below), has a lower cortex.

280b **Margins without cilia (a few projecting rhizines may be present in some species).**
.... **281**

281a (280) **Apothecia large, 2-3 mm in diameter, sunken in pits on the upper surface. Fig. 274.** ....
.... *Solorina saccata* (L.) Ach.

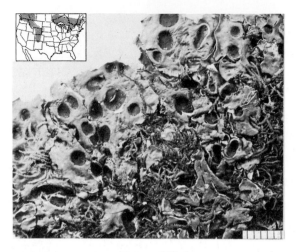

Figure 274

Figure 275

Figure 275   Physconia muscigena

Figure 274   Solorina saccata

Thallus greenish or brownish mineral gray, adnate, 3-6 cm broad; upper surface turning white pruinose; lower surface tan, without a cortex; apothecia common, the asci with 4 brown spores. Cortex and medulla K−, C−, P− (no substances). Rare on soil or over mosses in open woods or along streams. Two nearly identical species in western North America are separated by spore number: *S. bispora* Nyl. has two large brown spores and *S. octospora* Arn. has eight. All western specimens in this genus should be sectioned and examined with a microscope.

281b Apothecia smaller, adnate on the upper surface (on lower surface only in *Nephroma*), or apothecia lacking. 282

**282a (281) Upper surface scabrid and often becoming white pruinose (use hand lens and see Fig. 6D). Fig. 275. ....
.... *Physconia muscigena* (Ach.) Poelt**

Thallus light brownish gray to brown, extremely variable, taking on a white cast if heavily pruinose, adnate, 4-10 cm broad; upper surface with slightly raised margins and sparse lobules with age; lower surface brown to black, sparsely to densely rhizinate; apothecia not common. Cortex and medulla K−, C−, P−. Common over mosses on soil and over rocks, especially limestone, in open areas.

282b Upper surface usually shiny, smooth to ridged, not pruinose. 283

**283a (282) Lower surface bare, light brown; lobes broad and rotund, 2-6 mm wide, with isidiate-dentate margins. ....
.... (p. 134) *Nephroma helveticum***

283b Lower surface rhizinate, usually dark brown to black (buff only in *Anaptychia palmatula*); lobes generally narrow and linear, 0.5-3 mm wide. 284

284a (283) Thallus dark chestnut brown. .. 285

284b Thallus greenish to light brown to mineral gray. 287

285a (284) Margins of lobes with erect pycnidia (use hand lens and see Fig. 5C). Fig. 276. .... *Cetraria hepatizon* (Ach.) Vain.

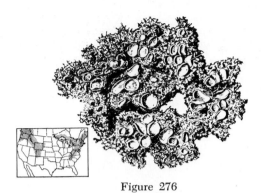

Figure 276

Figure 276   Cetraria hepatizon (×1.5)

Thallus dark brown, adnate, 4-8 cm broad; upper surface shiny, with conspicuous pycnidia; lower surface sparsely rhizinate; apothecia not common. Medulla K+ yellow→red, P+ orange (stictic acid). On exposed hard rocks in boreal regions or at higher elevations southward. *Parmelia stygia* is very close but has laminal pycnidia and reacts K− in the medulla.

285b Margins smooth; pycnidia (if present) laminal and immersed. 286

286a (285) Collected in boreal areas of northern United States and Canada. Fig. 277. .... *Parmelia stygia* (L.) Ach.

Figure 277

Figure 277   Parmelia stygia

Thallus dark brown, adnate, 4-7 cm broad; upper surface shiny; lower surface dark brown to black, sparsely rhizinate; apothecia not common. Medulla K−, C−, P+ red (protocetraric and fumarprotocetraric acids) or P− (fatty acids). Widespread in boreal or arctic regions on exposed rocks. *Hypogymnia oroarctica* occupies similar habitats and can be differentiated by the inflated lobes without rhizines. *Parmelia substygia* very rarely lacks soredia and would key out here. It can be separated by a C+ red, P− medulla test (gyrophoric acid).

286b Collected in southwestern United States. Fig. 278. .... *Neofuscelia atticoides* (Essl.) Essl.

Figure 278

Figure 278   Neofuscelia atticoides

Thallus chestnut brown, closely adnate on rock, 2-3 cm broad; lobes about 1 mm wide, becoming somewhat lobulate toward the center; pseudocyphellae completely lacking; lower surface tan, moderately rhizinate; apothecia not common. Medulla K+ yellow, C−, P+ orange (stictic and norstictic acids). Widespread on rock outcrops at higher elevations in Arizona and New Mexico. This genus, formerly considered in *Parmelia*, includes several other nonisidiate members that occur in the Southwest but are only rarely collected. One, *N. occidentalis* (Essl.) Essl., is P+ red (fumarprotocetraric acid). The others, *N. ahtii* (Essl.) Essl., *N. brunella* (Essl.) Essl., and *N. infrapallida* (Essl.) Essl., all contain fatty acids (medulla K−, P−), and should be identified with Esslinger's monograph.

287a (284) Upper surface reticulately ridged and with white markings (use hand lens and see Fig. 6C). .... (p. 97) *Parmelia omphalodes*

287b Upper surface uniform, without white markings. **288**

288a (287) Margins of lobes entire, without lobules. **289**

288b Margins of lobes lobulate. **290**

289a (288) Thallus closely adnate; lobes 0.5-1 mm wide; lower surface black. Fig. 279. .... *Phaeophyscia decolor* (Kashiw.) Essl.

Figure 279

Figure 279   Phaeophyscia decolor

Thallus dark mineral gray to faintly brownish, adnate on rocks, 2-5 cm broad; lower surface densely rhizinate; apothecia common. On rocks in fairly exposed areas at higher elevations. The medulla is white and contains zeorin. Plants with the red pigment skyrin in the medulla, rarely collected in Arizona, are called *P. endococcinodes* (Poelt) Essl.

289b Thallus adnate to loosely attached; lobes 1-1.5 mm wide; lower surface tan. Fig. 280. .... *Phaeophyscia constipata* (Norrl.) Moberg

Figure 280

Figure 280   Phaeophyscia constipata (×2)

Thallus light brownish gray, adnate to loosely attached, 3-7 cm broad; lower surface tan to buff, marginal rhizines sometimes projecting out from upturned lobe tips; apothecia rare. Cortex and medulla K–, C–, P– (no substances). Widespread on soil or among rocks in open areas. This species might be confused with *Anaptychia setifera* (see page 143) which has long cilia and lacks a lower cortex.

290a (288) Lobules not ascending. ....
.... **(p. 141)** *Anaptychia palmatula*

290b Lobes erect along margins. ....
.... **(p. 141)** *Phaeophyscia imbricata*

II. GELATINOUS LICHENS

These lichens contain blue-green algae, usually *Nostoc,* scattered through the heavily gelatinized thallus which swells when wet. There is little or no internal differentiation (see Fig. 9) and for this reason the medullary area is dark. This group as a whole prefers moist shady habitats although some species are typically found on exposed limestone or on soil.

Two major genera, *Collema* and *Leptogium,* make up most of the gelatinous lichens. They are sometimes very difficult to separate on external characters alone. While *Collema* is usually nearly black and dull and *Leptogium* is more bluish lead or slate-colored and shiny, some species of *Leptogium* are quite dark and a few Collemas are greenish brown. When in doubt, the only positive test is to make a thin vertical section with a razor blade and examine it under a microscope: *Leptogium* has a single layer of cells on the upper and lower surfaces while *Collema* shows no layering at all. Spore characters are important and sometimes indispensable for identifying species in both genera. The other important diagnostic characters are isidia (no soredia are produced), pustules, tomentum, and wrinkles. None of these lichens, unfortunately, has any chemical reactions and no color tests can be made on them.

1a   Collected submerged in mountain streams. Fig. 281. ....
.... *Hydrothyria venosa* Russell

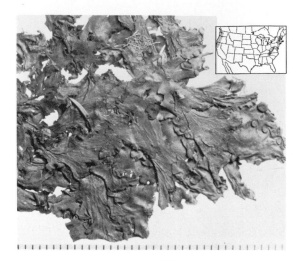

Figure 281

Figure 281   Hydrothyria venosa

Thallus bluish to dark mineral gray, growing in tufts 2-6 cm broad on rocks, leathery when dry; lower surface veined; apothecia common. Widespread, but rarely collected, in mountainous areas of the Appalachians and the Cascades. This is one of a few truly aquatic lichens, almost appearing to be a small brown alga. *Leptogium rivale* Tuck., known in the western states, is semi-aquatic near streams but smaller, less than 2 cm broad.

1b   Collected on trees, rocks, mosses, or soil. **2**

2a   (1) Lower surface whitish tomentose (use hand lens and compare with Fig. 7). **3**

2b   Lower surface smooth or wrinkled, lacking tomentum (tufts of white rhizine-like hapters may be present in some species of both *Collema* and *Leptogium*), nearly the same color as the upper surface to somewhat paler (or specimen too small to determine). **4**

3a   (2) Isidia or granules present on the upper surface; apothecia rare. Fig. 282. .... ***Leptogium burnetiae*** **Dodge**

Figure 282

Figure 282   Leptogium burnetiae (×1)

Thallus deep mouse gray, loosely adnate, 3-10 cm broad; lobes 3-10 mm wide, rotund; upper surface moderately isidiate with fine colorless hairs (use lens) near the margins; apothecia very rare. Common on trees and rocks in open woods. A closely related species more common in Canada is *L. saturninum* (Dicks.) Nyl., which is darker and has granular isidia. A smaller species, *L. laceroides* (Lesd.) Jørg., with lobes 1-3 mm wide and very short tomentum of sphaerical cells, occurs in the Appalachians. Another isidiate tomentose species, *L. papillosum* (Lesd.) Dodge, is found only in the southwestern states; it is strongly wrinkled.

3b   Isidia lacking; apothecia common. Fig. 283. .... ***Leptogium burgessii*** **(L.) Mont.**

Figure 283

Figure 283   Leptogium burgessii

Thallus bluish gray to brown, loosely adnate, 3-5 cm broad; upper surface plane to undulating; apothecia common, the rim lobulate, the spores muriform. On trees or on mosses over rocks in the southwestern states. Two other nonisidiate tomentose species will be collected in the Southwest. One, *L. digitatum* (Mass.) Zahlbr., has very short tomentum. The other, *L. rugosum* Sierk, lacks lobules on the apothecia, has long tomentum, and has elongate, 3-4 septate spores. It is deeply wrinkled as in *L. papillosum*.

4a   (2) Collected on rocks, soil, on mosses over soil, or more rarely at the base of trees. 5

4b   Collected on tree bark (also very rarely also on rock). 14

5a   (4) Upper surface isidiate or the lobe tips finely dissected and appearing isidiate. 6

5b   Upper surface smooth to wrinkled, lacking isidia. 10

6a   (5) Thallus bluish or brownish slate colored to dark mineral gray (see Frontispiece No. 13); upper surface shiny (use hand lens), smooth to finely wrinkled. 7

6b   Thallus brownish to greenish black (see Frontispiece No. 14); upper surface dull (use lens), usually smooth or pustulate but never wrinkled. 8

7a   (6) Lobes broad and rotund; isidia laminal. .... (p. 154) *Leptogium cyanescens*

7b   Lobes narrow, 1-2 mm wide, apically dissected to isidiate. Fig. 284. .... *Leptogium lichenoides* (L.) Zahlbr.

Figure 284

Figure 284   Leptogium lichenoides

Thallus dark brown, often pulvinate among mosses, 4-8 cm broad; upper surface finely wrinkled; apothecia lacking. Widespread on humus and mosses over rocks in fairly open areas and on rotting logs. This species is often overlooked because it blends with the substratum. *L. californicum* Tuck., a western species, is larger with less finely dissected lobes and a smooth cortex. Another similar species, *L. tenuissimum* (Dicks.) Fr., has smaller lobes with coralloid branching; it grows on sandy soil in much of the United States.

8a   (6) **Thallus appearing crustose, appressed on rock; lobes about 0.3 mm wide, with dense globular isidia and warts. Fig. 285.** ....
.... *Placynthium nigrum* (Huds.) S. Gray

Figure 285

Figure 285   Placynthium nigrum (×12)

Thallus brownish black, closely appressed on rock, 1-6 cm broad; lobes 0.2-0.3 mm wide, divided, with numerous isidia and isidia-like lobules; lower surface short bluish black tomentose, the tomentum projecting out as a mat around the thallus margin; apothecia common. Very common on limestone in open areas. This

lichen is typical of a large group of poorly known "cyanophilic" (with blue-green algal symbionts) species. They are all dark, scurfy to subcrustose, and usually grow on limestone. Henssen's articles should be consulted for more details.

8b   **Thallus distinctly foliose, adnate; lobes 2-5 mm wide.** **9**

9a   (8) **Lobes to 5 mm broad, more or less pustulate; isidia becoming lobulate. Fig. 286.** ....
.... *Collema flaccidum* (Ach.) Ach.

Figure 286

Figure 286   Collema flaccidum

Thallus olive green to dark brown, adnate, 4-8 cm broad; upper surface finely isidiate, becoming lobulate-isidiate with age; lower surface naked, wrinkled, greenish or olive gray; apothecia not common, the spores transversely septate. On limey rocks in fairly sheltered areas, rarely on trees. Other isidiate saxicolous or soil-inhabiting species, all of them very difficult to separate, include *C. crispum* (Huds.) Wigg. with 4-celled spores and *C. cristatum*

(L.) Wigg. with submuriform spores. These are found mostly in the western states.

9b Lobes 2-3 mm wide, not pustulate (or only weakly so); isidia warty, papillose or cylindrical. Fig. 287. ....
.... *Collema tuniforme* (Ach.) Ach.

11a (10) **Lobes broad and rotund; apothecia common.** ....
.... (p. 156) *Leptogium corticola*

11b Lobes narrower, elongate; apothecia rare. Fig. 288. ....
.... *Leptogium palmatum* (Huds.) Mont.

Figure 287

Figure 287   Collema tuniforme

Figure 288

Figure 288   Leptogium palmatum (×1)

Thallus black, rather closely adnate on rocks, 1-4 cm broad; lobes becoming crowded with ascending margins; surface plane to faintly pustulate, becoming densely isidiate, the isidia constricted at the base, simple to branched; lower surface bare or with sparse tufts of white hapters; apothecia rare; spores muriform. Widespread on exposed limestone. This is a commonly collected species and is not easily confused with any others. *Collema undulatum* Flot. var. *granulosum* Degel. is similar but with smaller, darker, more undulate lobes and 4-celled spores; it will be collected mostly in Canada.

Thallus bluish olive green to dark brown, loosely adnate on mosses on soil or over rocks, 5-8 cm broad; upper surface shiny, lobes becoming dissected, 100μ thick, slightly convoluted, sub-erect; lower surface dull, naked; apothecia common. Widespread over mosses in open areas. *Leptogium platynum* Herre, another western species, has a similar but much thicker thallus (more than 150 μm) *L. sinuatum* (Huds.) Mass., chiefly western but occurring sporadically eastward, has a smaller thallus with flattened lobe tips.

10a (5) **Thallus bluish slate-colored to dark mineral gray; surface shiny (use hand lens).** 11

10b Thallus brownish black to black; surface dull. 12

12a (10) **Lobes broad and rotund, more than 4-5 mm wide; surface pustulate.** Fig. 289.
.... *Collema ryssoleum* (Tuck.) Schneid.

Figure 289

Figure 289   Collema ryssoleum

Thallus olive or dark brown, adnate, 3-6 cm broad; lower surface olive greenish brown, smooth; apothecia very numerous, the spores 3-4 septate. Widespread on acidic rocks in eastern North America. This is the nonisidiate variant of *C. flaccidum*.

12b   Lobes narrow, crowded, 1-3 mm wide, often somewhat swollen and plicate. .. 13

13a   (12) Collected on rocks; lobules raised on edges. Fig. 290. .... *Collema polycarpon* **Hoffm.**

Figure 290

Figure 290   Collema polycarpon

Thallus olive green to dark brown, adnate, 2-5 cm broad; lobes swollen with a raised margin; lower surface olive green; apothecia numerous, the spores 1- or 2-septate. Widespread on limestone outcrops. In the same habitats one finds *Leptogium apalachense* Nyl., which has 3-5 septate muriform spores. Other confusable species include *Collema multipartitum* Sm., which has richly branched, less swollen, convex lobes. It is a western alpine species.

13b   Collected on soil. Fig. 291. .... *Collema tenax* (**Sw.**) **Ach.**

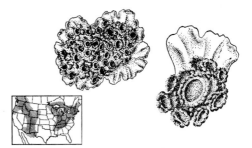

Figure 291

Figure 291 Collema tenax (×2)

Thallus blackish brown, adnate and forming a black crust, 3-6 cm broad; upper surface verrucose, the lobes irregular and crowded; apothecia common, the spores muriform, 20-25 μm long. Widespread on soil in open areas. This is the most common soil *Collema* but is usually overlooked. A minor variant in the western states (rarely Michigan and New York), *C. bachmanianum* (Fink) Degel., has a crenate rim and spores 30 μm long. Other soil species include *C. coccophorum* Tuck. with 1-septate spores and numerous erect lobules and *C. limosum* (Ach.) Ach., which has a subcrustose thallus and 4 muriform spores per ascus rather than 8. Serious students should consult Degelius' monograph to identify these species.

14a (4) **Thallus bluish to brownish slate or lead-colored; surface shiny (use hand lens), often finely wrinkled.** 15

14b **Thallus dull brownish black; surface not shiny (examine lobe tips), smooth to pustulate but never finely wrinkled.** .. 21

15a (14) Isidia present. 16

15b Isidia lacking. 19

16a (15) Isidia (and apothecia) located on strong ridges. Fig. 292. ....
.. *Leptogium marginellum* (Sw.) S. Gray

Figure 292

Figure 292 Leptogium marginellum

Thallus bluish gray, adnate, 4-6 cm broad; upper surface finely wrinkled; margins sparsely to moderately isidiate, the isidia simple, often associated with numerous apothecia with tan rim. Fairly common on deciduous trees in open woods or along roadsides. *Leptogium isidiosellum* (Ridd.) Sierk, a species occurring only in Florida, has cylindrical branched isidia along the ridges and usually lacks apothecia.

16b **Isidia produced over much of the thallus, not located on ridges.** 17

17a (16) **Lobes crowded and fusing together, forming an irregular network; surface very strongly wrinkled. Fig. 293.**
.... *Leptogium chloromelum* (Ach.) Nyl.

Figure 293

Figure 294

Figure 294 Leptogium cyanescens

Figure 293 Leptogium chloromelum

Thallus dark brownish gray, adnate on bark, 3-7 cm broad; isidia densely produced on major folds of the lobes; lower surface wrinkled; apothecia not common. Commonly collected on deciduous trees in open woods. If isidia are sparsely produced and overlooked, a specimen would be identified as *L. phyllocarpum,* but the apothecia of *L. chloromelum* lack the conspicuous lobules.

17b Lobes separate, not fusing together; surface smooth to finely wrinkled. 18

**18a (17) Lobe surface smooth; isidia fine, cylindrical, mostly laminal. Frontispiece No. 13 and Fig. 294. ....
...... *Leptogium cyanescens* (Ach.) Korb.**

Thallus bluish slate-colored, adnate, 3-7 cm broad; upper surface densely isidiate; apothecia very rare. This is probably the most commonly collected *Leptogium* in the deciduous forests of North America, but it can easily be confused in the South with *L. austroamericanum,* which has a wrinkled upper cortex. A species more common on limestone, *L. dactylinum* Tuck., is similar but has a smaller olivaceous thallus and numerous apothecia. A species with squamiform isidia, *L. denticulatum* Nyl., occurs on rocks from Colorado southward in the western states.

18b Lobe surface finely wrinkled (use hand lens); isidia coarse, becoming lobulate, in part marginal. Fig. 295. .... *Leptogium austroamericanum* (Malme) Dodge

Figure 295

Figure 295　Leptogium austroamericanum

Thallus bluish gray, adnate, 2-6 cm broad, often crowded; upper surface densely isidiate, dull; apothecia very rare. Common on deciduous trees in open dry woods. Within its range, this species is as common as *L. cyanescens* but is distinct because of the thicker, wrinkled thallus. A western species, *L. arsenei* Sierk, has a very thick thallus, 200-500 μm. *L. millegranum* Sierk has fused overlapping lobes and closely resembles *L. chloromelum* except for isidia; it has widespread occurrence in temperate North America.

19a　(15) **Lobes crowded and in part fusing, and forming an irregular network; upper surface strongly wrinkled; rim of apothecia conspicuously thickened and lobulate. Fig. 296.** ....
Leptogium phyllocarpum (Pers.) Mont.

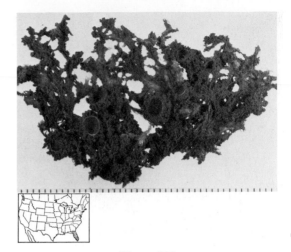

Figure 296

Figure 296　Leptogium phyllocarpum

Thallus dark bluish or brownish slate-colored, adnate, 3-8 cm broad; lobes 2-4 mm wide, fusing; upper surface with strong longitudinal wrinkles, the margins raised; apothecia large and conspicuous, the disk tan. Common on trees in open forests. This is a tropical species occurring mostly in Florida. A related species, *L. stipitatum* Vain., has apothecia borne on hollow inflated lobes. Another, *L. sessile* Vain., has orbicular rather than flattened lobes, and *L. floridanum* Sierk has thick warty lobes. Sierk's monograph should be consulted by students working in Florida.

19b　**Upper surface plane and smooth to weakly ridged; apothecia (if present) with a thin rim.** **20**

20a　(19) **Lobes more or less ascending, often marginally lobulate; apothecia rather rare. Fig. 297.** .... *Leptogium azureum* (Sw.) Mont.

Figure 297

Figure 297   Leptogium azureum

Thallus light bluish lead-colored, adnate, 4-8 cm broad; upper surface shiny; lower surface bluish gray; apothecia common. Rather rare on bark of trees in closed woods. *Leptogium cyanescens* is very close in color and habit but has isidia; *L. microstictum* Vain., known only from Florida, has a pitted upper surface. A rarely collected species in eastern North America, *L. crenatellum* Tuck., grows on trees near swamps where it is periodically inundated.

20b   **Lobes adnate, not lobulate; apothecia common. Fig. 298.** ....
.... ***Leptogium corticola* Tayl.**

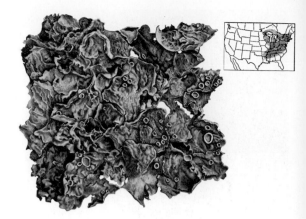

Figure 298

Figure 298   Leptogium corticola (×1)

Thallus bluish slate-colored, adnate, 3-8 cm broad; lobes 3-5 mm wide, rotund; lower surface slate gray; apothecia numerous on the upper surface. Widespread at the base of deciduous trees and rarely on rocks in moist woods. A rarer species in central United States that normally grows on rocks is superficially similar: *L. juniperinum* Tuck., which is smaller, darker, and has crowded apothecia.

21a   **(14) Thallus 2-6 cm broad; lobes expanded, separate, 2-5 mm wide; apothecia rare.** **22**

21b   **Thallus compact, 1-3 cm broad; lobes crowded and in part fused and swollen, 1-2 mm wide; apothecia often present.** ....
.... **24**

22a   **(21) Isidia lacking, the surface smooth to ridged. Fig. 299.** ....
.... ***Collema nigrescens* (Huds.) DC.**

Figure 299

Figure 299   Collema nigrescens

Thallus dark brownish to black, closely adnate on bark, 3-6 cm broad; lower surface pitted, greenish brown; apothecia very common, the disk lacking any white pruina, the spores 6-13 septate. Widespread on deciduous trees in open woods or along roads. *Collema pulcellum* Ach. var. *leucopeplum* (Tuck.) Degel. is virtually identical except for the white pruinose disk, while var. *subnigrescens* (Degel.) Degel. has clavate, 5-6 septate spores. It should be noted, too, that some specimens of *C. nigrescens* will have sparse globular isidia. These must be carefully distinguished from *C. furfuraceum* (below) which has mostly cylindrical isidia. Intergradations are numerous.

22b   Isidia present on lobe surface or pustules. 23

23a   (22) Thallus surface more or less flat. Figs. 28 and 300. .... *Collema subflaccidum* Degel.

Figure 300

Figure 300   Collema subflaccidum (×1.5)

Thallus deep olive brown to black, adnate, 4-8 cm broad; upper surface plane to undulating, the isidia dense, globose; lower surface naked, light olive brown; apothecia rare, the spores transversely septate. Very common at the base of trees in open forests and along roadsides. This species and *C. furfuraceum* are the most commonly collected species in the genus.

23b   Thallus surface pustulate and ridged (use hand lens). Fig. 301. .... *Collema furfuraceum* (Arn.) DR.

Figure 301

Figure 301 Collema furfuraceum

Thallus brownish black, adnate on bark, 2-4 cm broad; lobes rotund, 3-5 mm wide; isidia simple, more or less cylindrical; lower surface greenish black, smooth; apothecia rare. Very common on deciduous trees in open forests and along roads. The only confusable species is *C. subflaccidum* (above) which has a nonpustulate surface. Many specimens intermediate between these two will be found. In addition, abnormally isidiate specimens of *C. nigrescens* (above) will prove troublesome; the isidia are more globular.

24a (21) **Thallus 2-3 cm broad, forming a cushion covered with small apothecia. Fig. 302.** ....
.... *Collema conglomeratum* **Hoffm.**

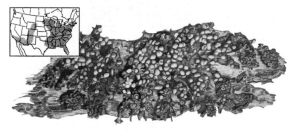

Figure 302

Figure 302 Collema conglomeratum (×1.5)

Thallus dark brown, closely adnate; lobes narrow and crowded; lower surface light colored; apothecia very abundant and crowded, the spores 1-septate. Common on deciduous trees

in open woods. Two very similar species must be examined microscopically for spores: *C. leptaleum* Tuck. has long multiseptate spores and *C. callibotrys* Tuck. has small muriform spores. Both occur most commonly in the southern states.

24b **Thallus 0.5-1 cm broad, appearing almost subcrustose; apothecia dispersed. Fig. 303. ..** *Collema fragrans* **(Sm.) Ach.**

Figure 303

Figure 303 Collema fragrans

Thallus dark olive green to brown or black, closely adnate with colonies coalescing; lobes more or less inflated, becoming suberect; apothecia common, the spores muriform. Widespread on deciduous trees in open woods but easily overlooked because of the small size. *Collema occultatum* Bagl. is similar but essentially subcrustose.

III. UMBILICATE LICHENS

Umbilicate lichens, the Rock Tripes, are attached to rocks by a single central cord below. They are basically a type of foliose lichen but without lobes or branches. Their favorite habitat is large boulders and cliffs. All of them are very brittle when dry and difficult to collect without fragmenting. When moistened they become soft and leathery and can be removed more easily.

The two most important genera, *Lasallia* and *Umbilicaria*, make up the family Umbilicariaceae. Some species of Lecanoraceae (*Rhizoplaca* and *Omphalodium*) and Verrucariaceae (*Dermatocarpon*) also have an umbilicate growth form and are included in this section, although taxonomically they are unrelated. The most important characters in *Lasallia* and *Umbilicaria* are isidia, soredia, and presence or absence of rhizines and plates on the lower surface. Rhizine characters are also important in *Dermatocarpon*. The differences between species are fairly sharp, but some specimens collected in exposed arctic-alpine habitats may be so modified that they do not key out well.

1a   **Thallus yellowish green (usnic acid present).** 2

1b   **Thallus whitish mineral gray, brown, or black.** 3

2a   **(1) Thallus 1-2 cm broad, closely adnate. Fig. 304.** .... *Rhizoplaca chrysoleuca* **(Sm.) Leuck. & Poelt**

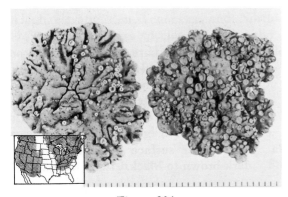

Figure 304

Figure 304   Rhizoplaca chrysoleuca

Thallus yellowish green to olivaceous, umbilicate; lower surface tan to brown, bare, with a thick central holdfast; apothecia very common, flesh-colored. Medulla K−, P− or P+ yellow (usnic acid, placodiolic acid, psoromic acid, or no substances). Common on exposed acidic rocks. The thalli are easily plucked off with a knife. In the western states a very similar species, *R. melanophthalma* (Ram.) Leuck. & Poelt, has dark rather than flesh-colored apothecia.

2b   **Thallus 3-8 cm broad, not closely attached. Fig. 305.** .... *Omphalodium arizonicum* **(Will.) Tuck.**

Figure 305

Figure 305   Omphalodium arizonicum (×1)

Thallus dull yellowish green, leathery; upper surface ridged, coarsely papillate; lower surface ridged, black to dark brown, sparsely rhizinate, the rhizines coarse, flattened; apothecia very common. On exposed rocks and cliff faces. This very unusual lichen is known at higher elevations in Arizona, New Mexico,

Figure 310

Figure 310 Umbilicaria vellea

Thallus 4-15 cm broad, leathery; upper surface scabrid or white pruinose; lower surface densely rhizinate, the rhizines coarse and simple; apothecia very rare. Medulla C+ red (gyrophoric acid). On exposed cliffs and boulders. A typical habitat for this relatively rare rock tripe is moist vertical cliffs. *U. mammulata* (above) intergrades with it but lacks pruina, has more delicate rhizines, and is brownish.

10a (6) Center of thallus with raised white ridges (see Fig. 316); rhizines very sparse. .... *Umbilicaria kraschennikovii* and *U. proboscidea* (see page 164)

10b Center of thallus uniform, flat, lacking ridges; rhizines moderate to dense. 11

11a (10) Lower surface with short dark rhizines; perithecia present on upper surface (use hand lens and see Fig. 309)..... .... (p. 164) *Dermatocarpon moulinsii*

11b Lower surface with long tan to pink or brown rhizines (or rhizines mixed with flat plates); perithecia lacking but apothecia often present. 12

12a (11) Rhizines dense and uniform, pinkish; plates lacking. Fig. 311. .... *Umbilicaria virginis* Schaer.

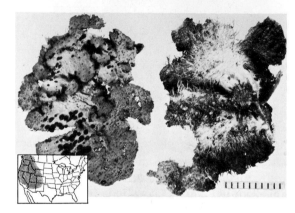

Figure 311

Figure 311 Umbilicaria virginis

Thallus whitish mineral gray, 2-6 cm broad; upper surface lightly scabrid and pruinose; lower surface densely rhizinate; apothecia very common, the disc plane, with a central fissure. Medulla C+ red (gyrophoric acid). Common on large boulders in more sheltered areas. *U. cylindrica* (L.) Del. is very close in appearance of the lower surface but has discs with concentric fissures. Both are typical arctic species.

12b Rhizines sparse, brown, intermingled with irregular flat plates (use hand lens and see Fig. 309). .... (p. 165) *Umbilicaria torrefacta*

13a (5) Upper surface strongly pustulate (do not use hand lens). Fig. 312. ....
.... *Lasallia papulosa* (Ach.) Llano

Figure 312

Figure 312   Lasallia papulosa

Thallus light brown, sometimes turning dull red, 3-15 cm broad, rather fragile; margins becoming lacerated with age; lower surface brown to tan, deeply pitted, bare; apothecia very common, black, the disc smooth. Medulla C+ rose (gyrophoric acid). Very common on exposed boulders and ledges. Another pustulate species, *L. pensylvanica* (Hoffm.) Llano, is jet black below; it is much rarer than *L. papulosa.*

13b   Upper surface plane to undulating or ridged, lacking pustules. 14

14a   (13) Upper surface with black dots (perithecia) (use hand lens and compare with Fig. 309); apothecia never present. 15

14b   Upper surface without black dots (perithecia) although some black pycnidia may be present; large black apothecia common. 16

15a   (14) Lower surface bare to very finely papillate. Fig. 313. ....
.. *Dermatocarpon miniatum* (L.) Mann.

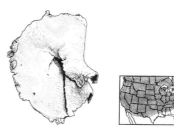

Figure 313

Figure 313   Dermatocarpon miniatum (×2)

Thallus pale brown to whitish gray, 2-5 cm broad, plane or becoming convoluted and crowded; lower surface dusky. Widespread on limestone and sandstone outcrops everywhere. This is an extremely variable lichen that can be confused with some Umbilicarias. There is a great deal of variation in crowding of the thallus, ranging from simple umbilicate plants to dense convoluted colonies that no longer appear umbilicate. In the western states it occurs with *D. moulinsii* (below) and *D. reticulatum* Magn. Near water on more acidic rocks it may be confused with *D. fluviatile* (see page 132) which turns green when wet and is more foliose. You will undoubtedly collect many specimens which do not seem to fit any one of these species but intergrade in the various characters.

15b   Lower surface with short rhizines. Fig. 314. .... *Dermatocarpon moulinsii* (Mont.) Zahlbr.

20a (18) Upper surface finely reticulately rugose (use hand lens and see Fig. 309). Fig. 319. ....
...... *Umbilicaria hyperborea* (Ach.) Ach.

Figure 319

Figure 319   Umbilicaria hyperborea

Thallus dark brown, 2-5 cm broad, the margins sometimes lacerated; lower surface brown, pitted; apothecia common, the disc with concentric fissures. Medulla C+ rose (gyrophoric acid). Widespread on exposed rocks at higher elevations. This is a common and variable species in the western states.

20b   Upper surface continuous, smooth. 21

21a (20) Apothecia common; thallus more or less round, adnate, the lower surface becoming finely papillate (see Fig. 309). Fig. 320. .... *Umbilicaria phaea* Tuck.

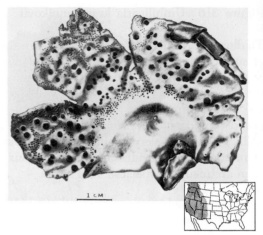

Figure 320

Figure 320   Umbilicaria phaea (×2)

Thallus brown, 2-3 cm broad; lower surface smooth to finely papillate, brown; apothecia very common, the disc concentrically fissured. Medulla C+ red, (gyrophoric acid). Very common on boulders and cliffs in open areas from southern California to British Columbia, rarer from New Mexico to Montana. This appears to be the commonest *Umbilicaria* on the West Coast.

21b   Apothecia rare; thallus dissected, irregularly ascending. Fig. 321. ....
.... *Umbilicaria polyphylla* (L.) Baumg.

Figure 321

Figure 321 Umbilicaria polyphylla

Thallus dark brown, leathery, 2-3 cm broad, dissected and appearing foliose; upper surface smooth; lower surface black, bare; apothecia rare. Medulla C+ rose (gyrophoric acid). On rocks in open talus slopes and ledges. The lobed, ascending thallus is unusual among the Umbilicarias.

IV. FRUTICOSE LICHENS

Fruticose lichens form a very large group of unrelated species that share a shrubby, hairlike, or strap-shaped growth form. The typical thallus consists of a main branch with or without numerous side branches. In *Cladina, Cladonia, Dactylina, Fistulariella, Pycnothelia,* and *Thamnolia* the branches are hollow but in the remaining genera they are more or less filled with hyphae. Rhizines are lacking, and basally attached species have instead a kind of holdfast at the base.

The genus *Cladonia* is by far the most widespread fruticose lichen and will be collected very frequently, even in cities. A specialized terminology has built up around *Cladonia.* For example, the small scalelike squamules, crowded into mats, form the primary thallus (Fig. 11). When only these squamules are present, it is often difficult to identify them to species (see Key V to the squamulose lichens). To make species identification one needs to collect the podetia, erect, simple or branched hollow structures borne on the squamules and often tipped with brown or red apothecia (see Fig. 10C). These podetia differ in shape, color,

axil openings, and presence of soredia (Fig. 337). The Reindeer Mosses (*Cladina* and *Cetraria*) lack the primary squamules and have highly developed, branched podetia or branches. Thomson (1967) has written a technical summary of *Cladonia* in North America with illustrations and notes on chemistry. This book should be consulted by serious students along with regional treatments by Brodo, Evans, Moore, and Wetmore, listed in the chapter on references.

Usnea, another commonly collected genus, is characterized by the yellow, richly branched thallus with a central cord. It is easily recognized in the field, but the species taxonomy can best be described as confused at this time. The diagnostic characters, soredia, isidia, and papillae, are extremely variable. The same can be said for *Ramalina,* which has not been revised in North America in recent times. *Bryoria,* a segregate of *Alectoria,* is also a difficult genus, although a monograph by Brodo and Hawksworth has cleared up many problems. *Stereocaulon* is another extremely plastic genus with many intergrading characters that require subjective decisions at the species level. In all of these cases the genera are easy to tell apart, but one should not be discouraged if some specimens do not seem to key out well.

The genera included in this section are listed below with brief diagnostic characters and the key couplet where most of the species are keyed.

Agrestia (85): fruticose, solid, soil-inhabiting; spores colorless, simple, 8/ascus.

Alectoria (109, 116): fruticose, solid, yellow (usnic acid), pseudocyphellate; spores brown, simple, 8/ascus.

Baeomyces (155): stalked with a primary thallus, solid; spores colorless, 1-3 septate, 8/ascus.

Bryoria (9): fruticose, brown; spores colorless, simple, 8/ascus (formerly in *Alectoria*).

Cetraria (4, 79): fruticose, flattened, pseudocyphellate, free growing; spores colorless, simple, 8/ascus.

Cladina (81, 146): fruticose with podetia, free growing, hollow, ecorticate; spores colorless, simple, 8/ascus (formerly in *Cladonia*).

Cladonia (20, 84, 87): fruticose with podetia and primary squamulose thallus, hollow, corticate; spores colorless, simple, 8/ascus.

Coelocaulon (5): fruticose, round, brown, free growing; spores colorless, simple, 8/ascus (formerly in *Cornicularia*).

Coenogonium (113): fruticose, hairlike, alga *Trentepohlia* dominating; spores colorless, simple or 1-septate, 8/ascus.

Cornicularia (7): fruticose, round, brown, usually free growing; spores colorless, simple, 8/ascus.

Dactylina (86): fruticose, round and hollow, free growing; spores colorless, simple, 8/ascus.

Dendrographa (130): fruticose, flattened, solid; apothecia ascolocular; spores colorless, 3-septate, 8/ascus.

Ephebe (6): fruticose, black, alga *Stigonema* dominating; spores colorless, simple, 8/ascus.

Fistulariella (95, 106): fruticose, flattened, hollow and perforate, yellow (usnic acid); spores colorless, 1-septate, 8/ascus (formerly in *Ramalina*).

Leprocaulon (137): fruticose with weakly differ-

entiated podetia; apothecia unknown (formerly in *Stereocaulon*).

Letharia (77): fruticose, flattened with some internal strands in the medulla, yellow (vulpinic acid); spores colorless, simple, 8/ascus.

Niebla (101): fruticose, flattened, yellow (usnic acid), the cortex palisade-paraplectenchymatous; spores colorless, 1-septate, 8/ascus (formerly in *Ramalina*).

Pilophoron (154): stalked, solid, with a primary thallus; spores colorless, simple, 8/ascus.

Pseudephebe (7): fruticose, round, solid, brown; spores colorless, simple, 8/ascus (formerly in *Alectoria*).

Pycnothelia (151): fruticose with podetia and a crustose primary thallus, hollow; spores colorless, simple, 8/ascus (formerly in *Cladonia*).

Ramalina (96): fruticose, flattened, pseudocyphellate, cortex prosoplectenchymatous, yellow (usnic acid); spores colorless, 1-septate, 8/ascus.

Roccella (129): fruticose, flattened and solid; apothecia ascolocular; spores colorless, 3-septate, 8/ascus.

Schizopelte (153): fruticose, rounded and solid; apothecia ascolocular; spores brown, 3-septate, 8/ascus.

Sphaerophorus (149): fruticose, round, solid; apothecia disintegrating at maturity; spores brown, simple, 8/ascus.

Stereocaulon (138): fruticose with solid pseudopodetia and phyllocladia; spores colorless, 3-septate, 8/ascus.

Teloschistes (16): fruticose, orange (parietin), round to flattened; spores colorless, 2-celled and polarilocular, 8/ascus.

Thamnolia (152): fruticose, free growing, hollow; apothecia unknown.

Usnea (110): fruticose with a central cord (rarely hollow), yellow (usnic acid); spores colorless, simple, 8/ascus.

1a **Thallus dark brown, often resembling horse hair, usually growing on trees (if growing on rocks or soil, lacking any squamules).** 2

1b Thallus orange, yellow, yellow green, reddish, white or mineral to brownish gray, growing on trees, rocks, or soil. .. 15

2a (1) Growing on soil, humus, or rocks. .. 3

2b Growing on trees. 8

3a (2) Branches flattened, large, 2-5 mm wide, becoming convoluted. 4

3b Branches round in cross section, narrow, 1 mm wide or less. 5

4a (3) Surface of thallus with large white pseudocyphellae, 1-3 mm wide; medulla P+ red. Fig. 322. Iceland Moss. ....
.... *Cetraria islandica* (L.) Ach.

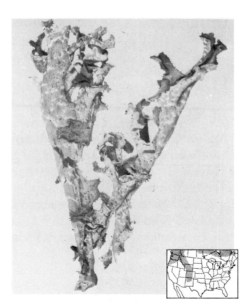

Figure 322

Figure 322   Cetraria islandica

Thallus loosely adnate in tufts 4-6 cm high, light brown on the outer side and tan on the inner side; lobes to 5 mm wide, usually convoluted, the margins with short spinules; base often red; apothecia not common. Medulla K−, P+ red (fumarprotocetraric acid and fatty acids). Common on soil and among mosses in arctic-alpine areas. This is essentially an arctic species, whereas *C. arenaria* (below) will be collected in temperate areas.

4b Pseudocyphellae inconspicuous or lacking; medulla P−. Fig. 323. ....
.... *Cetraria arenaria* Kärnef.

Figure 323

Figure 323   Cetraria arenaria (×2)

Thallus light greenish brown on the outer side, tan on the inner side, forming colonies

5-10 cm broad and 3-5 cm high on soil; lobes 1-5 mm wide, convoluted, the margins densely spinulate; apothecia rare. Medulla K−, C−, P− (fatty acids). Common on soil in fields or open pine forests. This species is disappearing as the open sandy areas where it flourishes are urbanized.

5a (3) **Collected in loose tufts on soil or humus. Fig. 324.** ....
Coelocaulon aculeatum **(Schreb.) Gyel.**

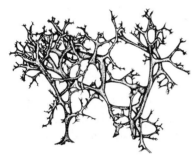

Figure 324

Figure 324   Coelocaulon aculeatum (×1)

Thallus chestnut brown, erect on soil or humus, 3-6 cm tall; branches shiny, sparsely pitted and perforate, very brittle; apothecia rare. Medulla K−, C−, P− (fatty acids). Widespread in sheltered areas in arctic or alpine localities. The spinulate branches distinguish it from *Cornicularia divergens* Ach., a larger species which reacts C+ red (olivetoric acid), and from any arctic Bryorias growing on soil.

5b **Collected growing attached to rocks.** .. 6

6a (5) **Thallus consisting of fragile black tufts less than 1 cm long; branches hair-like, about 0.1 mm thick. Fig. 325.** ....
.... ***Ephebe lanata*** **(L.) Vain.**

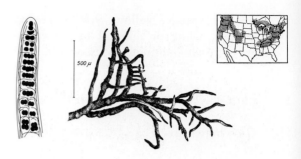

Figure 325

Figure 325   Ephebe lanata

Thallus black, dull, tufted and hair-like, flaccid, irregularly branched, branches up to 1 cm long and 70-140 μm thick, forming extensive mats on rock; hyphal cells elongate, parallel, the algal component *Stigonema;* apothecia rare. Widespread on moist rocks, especially near waterfalls. The algae make up most of the thallus. *Ephebe solida* Born., a second widespread species, is larger (up to 2 cm), stiff, and dichotomously branched, and has branches 130-260 μm thick. *E. americana* Henss. has thin branches (40-55 μm) but hyphae with angular cells arranged in irregular nets. All must be examined under the microscope.

6b **Thallus larger, 1-5 cm broad; branches more than 0.1 mm thick.** 7

7a (6) **Thallus prostrate on rock, richly branched. Fig. 326.** ....
...... ***Pseudephebe pubescens*** **(L.) Choisy**

Figure 326

Figure 326   Pseudephebe pubescens

Thallus dark brown to black, prostrate but loosely adnate, 3-5 cm broad; branches fine, round, 0.1-0.3 mm in diameter; medulla solid; apothecia rare. Medulla K−, C−, P− (no substances). Common on exposed boulders and ledges in open areas. It is probably most common in the Rocky Mountains. Another species, *P. minuscula* (Arn.) Brodo & Hawks., is similar but much smaller and almost subcrustose at the center with unevenly compressed branches and numerous apothecia. It occurs with *P. pubescens* and intergrades with it. Both species were formerly classified in *Alectoria*.

7b   Thallus suberect; branches 1 mm thick, sparsely branched. Fig. 327. ....
.. *Cornicularia normoerica* (Gunn.) DR.

Figure 327

Figure 327   Cornicularia normoerica (×1)

Thallus brownish black, suberect, stiff, 1-2 cm tall, sparingly branched; apothecia common, terminal, the rim dentate. Medulla K−, C−, P− (fatty acids). Rare, on exposed rocks. This curious lichen occurs only in the Cascades region from Oregon to British Columbia. It is probably more common than the few records indicate.

8a   (2) Thallus sorediate or sorediate-isidiate (use hand lens). 9

8b   Thallus without soredia or isidia, smooth.
... 11

9a   (8) Soredia in part isidiate or spinulate; thallus more or less prostrate. Fig. 328. *Bryoria furcellata* (Fr.) Brodo & Hawks.

Figure 328

Figure 328   Bryoria furcellata (×10)

Thallus chestnut brown, rather stiff, tufted to prostrate, 4-10 cm long; branches shiny; apothecia very rare. Medulla K−, P+ red (fumarprotocetraric acid). Very common on conifers and fenceposts in open areas. This is by far the commonest *Bryoria* in the eastern states but becomes quite rare in the West.

9b   Soredia powdery, without isidia intermixed; thallus usually pendulous or in part prostrate. 10

10a   (9) Soredia bright lemon yellow; thallus long pendulous. Fig. 329. .... *Bryoria fremontii* (Tuck.) Brodo & Hawks.

Figure 329

Figure 329   Bryoria fremontii

Thallus dark reddish brown to chestnut brown, 10-40 cm long, infrequently branched, uneven and twisted toward the base, to 0.4 mm in diameter; soralia rare, yellow (vulpinic acid); apothecia lacking. Medulla K−, C−, P− (no substances). Very common on conifers, especially top branches, in open forests. As discussed below, it frequently grows with *B. pseudofuscescens* and can be separated by the lack of pseudocyphellae and the stiffer, reddish hued thallus, as well as chemistry.

10b   Soredia white; thallus shorter, becoming prostrate in some habitats. Fig. 330. .... *Bryoria fuscescens* (Gyel.) Brodo & Hawks.

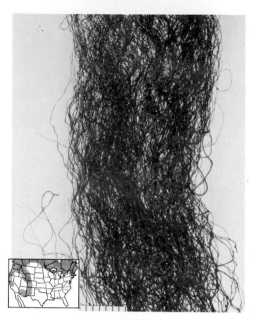

Figure 330

Figure 330   Bryoria fuscescens

Thallus dark brown, dull, 5-15 cm long, richly branched, the basal areas usually lighter brown; soralia conspicuous, wider than the branches; pseudocyphellae lacking; apothecia not found. Medulla and soralia K−, C−, P+ red (fumarprotocetraric acid). Very common on trunks of conifers in open woods. This is the commonest sorediate *Bryoria* in the western states. A very close species that occurs mostly in the Cascades, *B. glabra* (Mot.) Brodo & Hawks., is shiny and has smaller, oval soralia. Also on trees in the western states is *B. lanestris* (Ach.) Brodo & Hawks., which is extremely brittle and very dark and has narrower branches. Another superficially similar sorediate species, *B. friabilis* Brodo & Hawks., is distinguished by a C+ rose test (gyrophoric acid) and presence of pseudocyphellae. A soil and tundra species, *B. chalybeiformis* (L.) Brodo & Hawks., will also key out here. And, finally, there is *B. nadvornikiana* (Gyel.) Brodo & Hawks., which occurs in the Great Smokies, New England, and Canada, characterized by a pale brown thallus and K+ deep yellow reaction (diffuse on filter paper; alectorialic and barbatolic acids). The monograph by Brodo and Hawksworth should be consulted for more information on these difficult species.

11a **(8) Thallus uniformly pale brown to buff; K+ deep yellow (diffused on filter paper). Fig. 331. .... *Bryoria capillaris* (Ach.) Brodo & Hawks.**

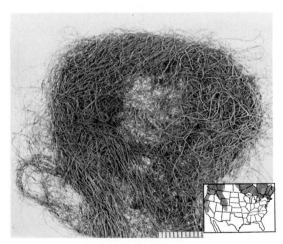

Figure 331

Figure 331   Bryoria capillaris

Thallus pendulous, 10-20 cm long, soft, draped over branches without basal attachment; branches fine, evenly thickened, 0.1-0.2 mm in diameter; apothecia rare. Cortex and medulla K+ deep yellow (alectorialic and barbatolic acids). Widespread and locally abundant on conifers. The K test is made by putting frag-

ments of thallus on filter paper and adding a drop of KOH. No other nonsorediate *Bryoria* will react K+ yellow.

11b **Thallus dark reddish to chestnut or blackish brown; K− or K+ yellow turning red. 12**

12a **(11) Thallus rather closely adnate and tufted on branches, 3-6 cm high; apothecia common. Fig. 332. .... *Bryoria abbreviata* (Müll. Arg.) Brodo & Hawks.**

Figure 332

Figure 332   Bryoria abbreviata

Thallus dark brown, irregularly branched and spinulate; apothecia common, the rim spinulate. Medulla K−, P− (fatty acids). Common on branches of conifers in exposed areas at higher elevations. The thallus branches are somewhat flattened. A companion species, *B. oregana* (Nyl.) Brodo & Hawks., lacks apothecia and is pendulous. *Cornicularia californica* may key here as well; it has irregularly flattened branches and apothecia with a smooth rim.

12b Thallus pendulous, 10-40 cm long; apothecia rare. 13

13a (12) Thallus light brown, soft, the branches very thin (0.1-0.2 mm in diameter); collected in eastern North America and the Cascades. Fig. 333. .. *Bryoria trichodes* (Ach.) Brodo & Hawks.

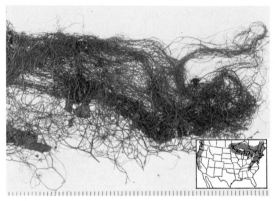

Figure 333

Figure 333   Bryoria trichodes

Thallus 10-30 cm long with very fine, smooth branches; apothecia rare. Medulla K−, C−, P+ red (fumarprotocetraric acid). Common on conifers in the boreal forest but rarer southward in the mountains. This is the main pendulous *Bryoria* that you are likely to collect in eastern North America. It was variously called "Alectoria jubata" or "A. chalybeiformis" in the old texts.

13b Thallus dark to reddish brown, stiffer; branches to 0.4 mm thick at the base, thinner at the tips; collected on western North America. 14

14a (13) Thallus dark brown; white pseudocyphellae fairly conspicuous (use hand lens); medulla K+ yellow turning red. Fig. 334. .... *Bryoria pseudofuscescens* (Gyel.) Brodo & Hawks.

Figure 334

Figure 334   Bryoria pseudofuscescens

Thallus pendulous, 5-25 cm long; surface of branches shiny; apothecia very rare. Medulla K+ yellow turning red, P+ orange (norstictic acid). Very common on conifers in open forests in the Northern Rockies and the Cascades. Superficially it is very similar to the equally common, nonsorediate forms of *B. fremontii*, which is K−, more reddish tinged, and generally more robust. It also lacks pseudocyphellae, although this may not be an easy character to determine. If possible, chemical tests should be made. Another species in this group, *B. tortuosa* (Merr.) Brodo & Hawks., occurs commonly in the Cascades; it has yellow pseudocyphellae (vulpinic acid present).

14b Thallus dark reddish brown; pseudocyphellae rare to absent; medulla K−. .... (p. 172) *Bryoria fremontii*

15a (1) Thallus and apothecia (if present) orange, reacting K+ purple. **16**

15b Thallus yellow, yellowish green, mineral to brownish gray, or white, K— or K+ yellow. **17**

16a (15) Thallus tufted, 1-2 cm high. Fig. 335. .... *Teloschistes chrysophthalmus* (L.) Th. Fr.

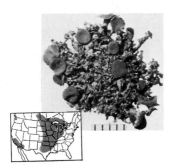

Figure 335

Figure 335   Teloschistes chrysophthalmus

Thallus pale orange to gray, tufted, 1-2 cm broad, loosely attached; branches flattened, little branched but becoming spinulate or ciliate marginally; apothecia very common, the rim spinulate. Widespread on exposed trees or on branches at the tops of trees. This colorful lichen is smaller than *T. exilis* and apparently more common.

16b Thallus loosely tufted to pendulous, 3-8 cm long. Fig. 336. .... *Teloschistes exilis* (Michx.) Vain.

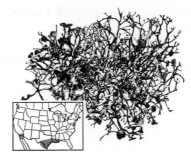

Figure 336

Figure 336   Teloschistes exilis (×1)

Thallus pale to deep orange, 3-7 cm broad, loosely attached; branches flattened to round, becoming finely spinulate at the tips; apothecia common. Common on exposed trees in prairie regions or open dry uplands. A closely related but rarer species, *T. flavicans* (Sw.) Ach., is sorediate and lacks apothecia.

17a (15) Thallus consisting of erect cup-shaped, pointed, or branched, hollow podetia arising from basal squamules (see Fig. 337); if basal squamules lacking, the podetia round in cross section, hollow, and sparsely to conspicuously squamulate; collected most often on soil but also growing on mosses and over mosses or humus on rocks and the base of trees. **18**

Figure 337

Figure 337 Examples of kinds of podetia in *Cladonia*

17b Thallus erect, pendulous, or prostrate, never cup-shaped or with basal squamules (*Stereocaulon* has small phyllocladia; see Fig. 443); branches solid or hollow, round or flattened; collected on trees and rocks, more rarely free growing on soil. 75

18a (17) Podetia forming distinct cups (see Fig. 337). 19

18b Podetia not cup-shaped but forming pointed or blunt clubs (see Fig. 337) often tipped with apothecia (if branched, irregular cups may be formed by the expanded axils). 38

19a (18) Apothecia (or sterile pycnidia) bright red (use hand lens for pycnidia), K+ purple. 20

19b Apothecia and pycnidia pale to dark brown, K− or K+ brownish; or apothecia and pycnidia lacking. 23

20a (19) Podetia coarsely areolate, without soredia. Fig. 338. ....
.... *Cladonia coccifera* (L.) Willd.

Figure 338

Figure 338 *Cladonia coccifera* (×1)

Cups yellowish green, stout, 1-2 cm tall, sometimes proliferating marginally, the surface covered with scattered areoles; primary squamules well developed at the base of podetia; pycnidia and apothecia very common. K−, C−, P− (barbatic and usnic acids). Widespread on soil and on humus over rocks in open alpine or arctic areas. Closely related *C. pleurota* has distinct granular soredia.

20b Podetia covered with powdery soredia. 21

21a (20) Podetia whitish gray, the surface K+ deep yellow. Fig. 339. *Cladonia digitata* (L.) Hoffm.

Figure 339

Figure 339 *Cladonia digitata* (×1)

Cups whitish to yellowish mineral gray, 1.5-4.0 cm high, rarely proliferating marginally, corticate toward the base but becoming densely sorediate toward the upper part; primary squamules large, sorediate; apothecia and pycnidia common, blood red. K+, P+ deep yellow (thamnolic acid). Rare on humus and over mosses in open areas or swamps.

21b Podetia yellow to greenish yellow, the surface K− or turning yellowish green. 22

22a (21) **Cups short and stout. Fig. 340.** ....
.... ***Cladonia pleurota*** (Flk.) Schaer.

Figure 340

Figure 340 Cladonia pleurota (×1)

Cups stout, 1-2 cm tall, rarely proliferating, sorediate in the upper parts and inside the cup; primary squamules distinct, medium-sized; pycnidia and apothecia common, deep red. Common on soil and on humus over rocks in the open. At first this distinctive cup *Cladonia* will be confused with greenish forms of the *C. chlorophaea* group until the red pycnidia are seen. *C. deformis* has much taller, narrow cups.

22b **Cups tall and narrow. Fig. 341.** ....
.... ***Cladonia deformis*** (L.) Hoffm.

Figure 341

Figure 341 Cladonia deformis (×1.5)

Cups yellowish green, 2-4 cm tall, irregularly expanded, diffusely sorediate but with corticate areas near the base; primary squamules medium-sized, sparsely developed; pycnidia common, apothecia rather rare, blood red. Widespread on humus and rotting logs in swamps and open areas. Generally only a few

podetia will be found at a time. *C. gonecha* (Ach.) Asah., is virtually indistinguishable except for larger primary squamules (5-10 mm long) and the presence of squamatic acid with or without bellidiflorin. Both of these have much taller, narrower cups than *C. pleurota*.

23a (19) **Podetia covered with powdery or granular soredia.** 24

23b **Podetia lacking soredia** (***Cladonia pyxidata*** has coarse areoles and *C. squamosa* fine dense squamules). 29

24a (23) **Cups deep, stout to elongate, distinct.** 25

24b **Cups shallow, poorly developed, podetia in part pointed or lacerated at the tips.** 28

25a (24) **Cups yellow** (usnic acid present). **Fig. 342.** ...... ***Cladonia carneola*** (Fr.) Fr.

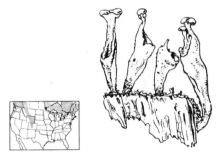

Figure 342

Figure 342 Cladonia carneola (×1.5)

Cups 1.5-3.0 cm high, sometimes proliferating marginally, diffusely sorediate; primary squamules small, incised, poorly developed; pycnidia common, apothecia rather rare, pale brown. Rare on humus and soil in open areas.

It resembles *C. chlorophaea* but has a distinctive yellowish cast and more extensive powdery soredia. The brown pycnidia and apothecia will separate it from redfruited *C. pleurota*.

25b **Cups white, greenish gray to faint brownish.** 26

26a **(25) Soredia coarse and granular; cups generally stout, mineral gray. Figs. 30 and 343.** ....
.... ***Cladonia chlorophaea* (Flk.) Spreng.**

Figure 343

Figure 343   Cladonia chlorophaea (×1.5)

Cups 0.5-1.5 cm high, greenish mineral gray, simple or proliferating marginally, the axils closed; soredia granular, quite diffuse to sparse, intergrading with areoles; primary squamules rather coarse, sparsely to moderately developed; apothecia rare, dark brown. Widespread and common on soil and over mosses on roadbanks, over rocks, and at bases of trees. This is the most commonly collected cup *Cladonia*. There are five chemical populations which can only be separated by microchemical crystal tests: *C. chlorophaea* (fumarprotocetraric acid), *C. grayi* Merr. (grayanic with or without fumarprotocetraric), *C. cryptochlorophaea* Asah. (cryptochlorophaeic with

or without fumarprotocetraric acid), *C. merochlorophaea* Asah. (merochlorophaeic with or without fumarprotocetraric), and *C. perlomera* Krist. (merochlorophaeic, 4-0-methylcryptochlorophaeic and, perlatolic acids). *Cladonia cryptochlorophaea*, *C. grayi*, and *C. perlomera* occur in the eastern states while *C. merochlorophaea* is boreal and western. *C. chlorophaea* is everywhere.

26b **Soredia fine and powdery; cups generally thinner.** 27

27a **(26) Cups deep and expanded. Fig. 344.** .... ***Cladonia conista* (Ach.) Robb.**

Figure 344

Figure 344   Cladonia conista (×1)

Cups whitish mineral gray, 0.5-1.0 cm tall; soredia farinose, diffuse; primary squamules coarse, numerous; apothecia very rare. P+ red (fumarprotocetraric acid) or P−. Widespread on soil and over mosses in open areas. *C. major* (Hag.) Sandst. is very similar except for being about twice as large and lacking substance H. Both intergrade with and must be carefully distinguished from *C. chlorophaea*.

27b **Cups tall, not expanded. Fig. 345.** ....
.... ***Cladonia fimbriata* (L.) Fr.**

Figure 345

Figure 345   Cladonia fimbriata (×1)

Cups whitish mineral gray, narrow, 1-3 cm high, rarely proliferating marginally; soredia farinose, diffuse; primary squamules small, sparsely developed; apothecia rare, dark brown. P+ red (fumarprotocetraric acid). Widespread on soil in open areas. This species intergrades with both *C. conista* and *C. chlorophaea* but is generally recognized by the narrower cups.

28a **(24) Cups lacerated at the tips, the margins rolled inward. Fig. 346.** ....
.... ***Cladonia cenotea*** **(Ach.) Schaer.**

Figure 346

Figure 346   Cladonia cenotea (×1.5)

Podetia whitish mineral gray, 1-4 cm tall, narrow, sparingly branched in the upper parts; becoming squamulate toward the base, soredia diffuse over most of the surface; primary squamules medium-sized; pycnidia common,

apothecia rare, dark brown. Widespread on rotten stumps and wood and on soil or humus in open areas. The asymmetrical cups with margins rolled inward are characteristic of this northern *Cladonia*.

28b **Cups entire, often very small, the margins flaring. Fig. 347.** ....
.... ***Cladonia rei*** **Schaer.**

Figure 347

Figure 347   Cladonia rei (×1)

Podetia whitish gray, 1-3 cm high, simple or sparingly branched, pointed or flaring into small irregular cups, the rim indented or proliferating; sparsely sorediate over much of the surface but with bare ecorticate areas showing, squamulate toward the base; primary squamules small; pycnidia common. Cortex P+ red (fumarprotocetraric acid) or P− (homosekikaic acid). On soil and humus in pastures or on roadbanks, often resembling dried grass stubble. This variable species may resemble *C. fimbriata* if cups are well developed or *C. decorticata* if cups are lacking.

29a **(23) Cups proliferating from the center.** **30**

29b **Cups proliferating from the margins or not proliferating at all.** **31**

30a (29) Cups relatively deep; collected in western and eastern North America. Fig. 348. .... *Cladonia verticillata* (Hoffm.) Schaer.

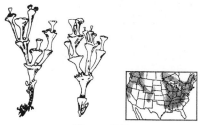

Figure 348

Figure 348   Cladonia verticillata (×1)

Podetia greenish mineral gray, 2-8 cm tall, smooth to moderately squamulate; primary squamules small, generally poorly developed; pycnidia and apothecia common, dark brown. P+ red (fumarprotocetraric acid). Widespread on soil in pastures and along roadsides. This species can be recognized at sight by the unusual verticillate cups. Toward the southern part of its range it occurs intermingled with *C. calycantha* Nyl.

30b Cups shallow and flaring; collected in the Coastal Plain from New England to Florida and Louisiana. Fig. 349. .... *Cladonia calycantha* Nyl.

Figure 349

Figure 349   Cladonia calycantha (×2)

Podetia rather delicate, whitish gray, growing in colonies on sandy soil, 1-5 cm tall; cups 3-5 mm wide, proliferating centrally in several tiers. Cortex K−, P+ red (fumarprotocetraric acid). Common in open woods or scrub land. *C. verticillata* is rare in the Coastal Plain where *C. calycantha* is so common. A chemical variant with psoromic acid (K−, P+ deep yellow), *C. rappii* Evans, will be collected in Florida.

31a (29) Cups coarse and stout, usually not proliferating, covered with coarse greenish areoles. Fig. 350. .... *Cladonia pyxidata* (L.) Hoffm.

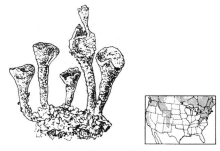

Figure 350

Figure 350   Cladonia pyxidata (×1.5)

Cups greenish mineral gray, 1.0-1.5 cm high, the surface very coarsely areolate and granular but not forming soredia; primary squamules often well developed, large; pycnidia common, apothecia rare, dark brown. P+ red (fumarprotocetraric acid). Common on humus and soil over rocks in open areas. When well developed, this is an easily recognized lichen but it intergrades with the *C. chlorophaea* group in the southern part of its range. The differences between small areoles and large granular soredia are often difficult to detect and one cannot avoid collecting specimens that seem to be intermediates.

31b Cups more attenuated, more or less pro-liferating, areoles not conspicuously de-veloped. 32

32a (31) Centers of cups closed (use hand lens). 33

32b Centers of cups open or perforated. 35

33a (32) Cortex lacking at the tips, the sur-face fibrous (use hand lens); base of podetia black-spotted. Fig. 351. ....
.. *Cladonia phyllophora* (Ehrh.) Hoffm.

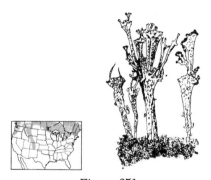

Figure 351

Figure 351   Cladonia phyllophora (×1)

Podetia whitish mineral gray, 3-5 cm tall, sparingly branched, the cups irregularly lacer-ate and partially obscured by squamules; pri-mary squamules medium-sized, well devel-oped; pycnidia and apothecia common, dark brown. P+ red (fumarprotocetraric acid). Rather rare on soil and over mosses in open areas. This species in its typical form is not usually confusable with other Cladonias that have squamulate irregularly cup-forming po-detia. However, it must be carefully distin-guished from *C. crispata* (K−) and *C. multi-formis* (cups regularly perforate).

33b Cortex continuous at tips, shiny; base uniform, not black-spotted. 34

34a (33) Surface of podetia K−, atranorin lacking. Fig. 352. ....
.... *Cladonia gracilis* (L.) Willd.

Figure 352

Figure 352   Cladonia gracilis (×1.5)

Podetia brownish mineral gray, 2-8 cm tall, sparingly branched, the cups quite distinct to abortive, smooth to moderately squamulate; primary squamules generally poorly devel-oped, often evanescent in the boreal forests; pycnidia and apothecia common, dark brown. P+ red (fumarprotocetraric acid). Common and widespread on soil, humus, and mosses in open areas. Probably no other *Cladonia* ex-hibits as much variation as this species, and there is not space to enumerate all of the vari-eties here. Basically it is recognized by the brownish tinge, marginally proliferating cups, and continuous shiny cortex.

34b Surface of podetia K+ yellow (atranorin present). Figs. 10B and 353. ....
.... *Cladonia ecmocyna* (Ach.) Nyl.

Figure 353

Figure 353   Cladonia ecmocyna (×2)

Podetia greenish gray, loosely attached on soil or among mosses with few basal squamules, rather coarse and sparsely branched and forming dense colonies to 15 cm broad; cups barely forming, apothecia usually present at tips; coarse squamules sparsely produced toward base of podetia. Cortex K+ yellow (atranorin), P+ red (fumarprotocetraric acid). Common in open conifer forests and along trails. There is much resemblance to some forms of *C. gracilis* which are blunt-tipped rather than pointed, and a chemical test should be made on all specimens.

35a   (32) **Membrane of cups perforated with small holes (use hand lens). Fig. 354.** .... *Cladonia multiformis* **Merr.**

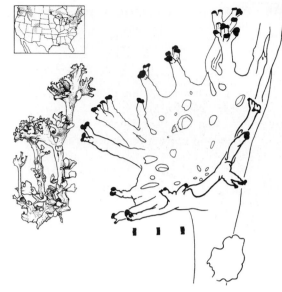

Figure 354

Figure 354   Cladonia multiformis (×1.5)

Podetia greenish to brownish mineral gray, 3-5 cm tall, sparsely to moderately squamulate, the cups proliferating marginally; primary squamules medium-sized, often sparsely developed; pycnidia very common, apothecia rare, dark brown. P+ red (fumarprotocetraric acid). Widespread on soil along road cuts and in open fields. This species is extremely variable in development of cups and density of squamules, but the perforate membrane is usually sufficiently distinct for positive identification. *C. crispata* (P−, squamatic acid) and *C. caraccensis* Vain. (K+ yellow thamnolic acid) are closely allied, rare species best distinguished by the chemical features.

35b   **Cups open and gaping.** 36

36a   (35) **Podetia finely and densely squamulose (use hand lens). Fig. 355.** .... *Cladonia squamosa* (**Scop.**) **Hoffm.**

Figure 355

Figure 355　Cladonia squamosa (×1)

Podetia greenish mineral gray, 3-6 cm tall, sparingly branched, forming irregular narrow cups, the axils open; primary squamules poorly developed or evanescent; pycnidia common, apothecia very rare. Widespread on soil and over mosses on rocks in mature forests. This common *Cladonia* may be difficult to identify at first because of the variability in cup development. It fluoresces brilliant white in UV. In the Pacific Northwest there is one chemical variant, *C. subsquamosa* (Nyl.) Vain. which is K+, P+ yellow (thamnolic acid).

36b　Podetia sparsely to moderately squamulose. **37**

37a　(36) Basal squamules large, 2-3 mm long. Fig. 356. .... *Cladonia mateocyatha* **Robb.**

Figure 356

Figure 356　Cladonia mateocyatha (×1.5)

Podetia dark greenish mineral gray, up to 1 cm tall but often aborted and lacerated, cup-shaped with a closed membrane; primary squamules well developed, coarse, up to 1 cm long, forming extensive mats, the lower surface cream or buff to tan; apothecia and pycnidia rare. P+ red (fumarprotocetraric acid). Widespread on soil in open woods, road cuts and soil banks. It often occurs with *C. apodocarpa,* which has more linear squamules with a chalky white lower surface. *C. turgida* is generally much larger and reacts K+ yellow (atranorin).

37b　Basal squamules small or lacking. Fig. 357. .... *Cladonia crispata* (Ach.) **Flot.**

Figure 357

Figure 357　Cladonia crispata (×1.5)

Podetia brownish mineral gray, 5-8 cm tall, shiny, branched with flaring cups, the cups proliferating marginally, the axils open, sparsely to moderately squamulose; primary squamules evanescent; apothecia and pycnidia common, small, dark brown. K−, P− (squamatic acid). On soil or over mosses in open areas. Two chemical variants are restricted to the southeastern states: *C. atlantica* Evans which contains squamatic and baeomycic acids (P+ yellow) and *C. floridana* Evans which contains thamnolic acid (K+ yellow).

38a (18) **Apothecia (if present) and pycnidia red (use hand lens), K+ purple. .. 39**

38b **Apothecia (if present) and pycnidia brown or black; or apothecia and pycnidia lacking.** 46

39a (38) **Podetia and squamules lacking soredia. Fig. 358. British Soldiers.** .... *Cladonia cristatella* **Tuck.**

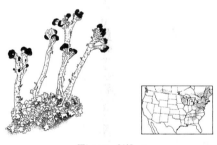

Figure 358

Figure 358 Cladonia cristatella (×1.5)

Podetia yellowish green, 1-2 cm tall, branched toward the upper parts, smooth to moderately squamulate; primary squamules inconspicuous; apothecia and pycnidia very common. Very common on humus, soil, and rotting logs in open areas. This is one of the first lichens collected by a lichen student. In Florida it is replaced by *C. abbreviatula* Merr., a smaller species that reacts K+ yellow (didymic and

thamnolic acids). In the Pacific Northwest and arctic Canada the common esorediate red-fruited species is *C. bellidiflora* (Ach.) Schaer., larger and more richly squamulate than *C. cristatella*.

39b **Podetia and/or squamules more or less covered with powdery soredia.** 40

40a (39) **Podetia pale yellow (usnic acid present); Pacific Northwest. Fig. 359.** .... *Cladonia transcendens* **(Vain.) Vain.**

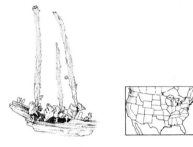

Figure 359

Figure 359 Cladonia transcendens (×1.5)

Podetia pale yellowish green to gray, 3-4 cm tall, sparsely to densely squamulate, sparingly branched and sometimes with narrow cups, the axils closed, sorediate toward the upper parts; primary squamules medium-sized to large, incised; pycnidia and apothecia common, red. K+, P+ yellow (thamnolic acid with or rarely without usnic acid). Widespread on stumps and decaying logs in open conifer forests. This is a western species separated from *C. bellidiflora* by the presence of soredia.

40b **Podetia whitish to greenish mineral gray (usnic acid lacking except in *Cladonia incrassata*); eastern North America (*Cladonia bacillaris* and *C. macilenta* rarely in the west).** 41

41a (40) Surface of podetia instantly K+ deep yellow (thamnolic acid present). 42

41b Surface of podetia K− or slowly dingy yellow or brown. 43

42a (41) Lower surface of squamules uniformly white. Fig. 360. .... *Cladonia macilenta* Hoffm.

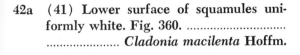

Figure 360

Figure 360   Cladonia macilenta

Podetia whitish mineral gray to white, 0.5-1.5 cm tall, unbranched to sparingly branched toward the apex; soredia farinose, diffuse; primary squamules sparse to dense, small; pycnidia common, apothecia rare, K+, P+ deep yellow orange (thamnolic acid). Common on fenceposts, shingles, and base of trees in the open. *C. bacillaris,* a more widespread species, differs only in chemistry (K−, P−). In the Coastal Plain from North Carolina through Florida to Texas one will find *C. ravenelii* Tuck., which has smaller podetia with many primary squamules and also reacts K+ yellow.

42b Lower surface of squamules with pale orange veins. Fig. 361. .... *Cladonia hypoxantha* Tuck.

Figure 361

Figure 361   Cladonia hypoxantha (×2)

Podetia whitish mineral gray, short and often poorly developed, 0.5-1.0 cm high, sorediate in the upper parts, inner cartilaginous layer pale orange; primary squamules well developed and forming extensive mats, sorediate, incised; pycnidia and apothecia rare, red. K+, P+ yellow (thamnolic acid). Widespread on humus and on rotting logs in open woods in Florida. Though at first appearing to be an indeterminate mass of squamules, this species is immediately recognized by the unique orange veins on the lower surface of the squamules.

43a (41) Primary squamules rather large 2-4 mm), sorediate. Fig. 362. .... *Cladonia incrassata* Flk.

Figure 362

Figure 362   Cladonia incrassata

Podetia greenish or yellowish mineral gray, 0.5-2.0 cm tall, smooth to sparsely squamulate; primary squamules becoming quite large and forming an extensive mat, powdery sorediate along the margins; apothecia common. On logs and base of trees in open woods or in swamps. *C. cristatella* is closely related but lacks any trace of soredia on the squamules.

43b   Primary squamules small and usually poorly developed, esorediate or with sparse soredia. 44

44a   (43) Podetia commonly more or less branched, robust; large corticate areas remaining. Fig. 363. .... *Cladonia floerkeana* (Fr.) Somm.

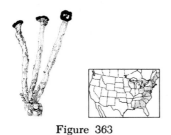

Figure 363

Figure 363   Cladonia floerkeana (×1.5)

Podetia whitish mineral gray, 0.5-2.5 cm tall, unbranched, smooth or covered with small squamules, sparsely to moderately sorediate but with distinct corticate areas toward the apex; primary squamules, sparse, tiny, turning sorediate; apothecia common. Widespread on soil, logs, and humus in open areas. This species may be confused with *C. cristatella,* a much more common lichen which lacks any soredia. *C. bacillaris* and *C. macilenta* have extensive diffuse soredia and usually lack apothecia.

44b   Podetia usually simple, often pointed, largely ecorticate. 45

45a   (44) Ecorticate areas bare and darkening. Fig. 364. .... *Cladonia didyma* (Fée) Vain.

Figure 364

Figure 364   Cladonia didyma (×2)

Podetia whitish mineral gray, 0.5-1.5 cm tall, mostly unbranched, sparsely to densely covered with granular or isidioid squamules, the ecorticate areas turning brownish and translucent; primary squamules usually dense, finely incised; apothecia common. Widespread on soil in open areas. This is probably the most common red-fruited *Cladonia* in the southeastern states. *C. bacillaris* is close but has farinose soredia and is usually larger.

45b   Ecorticate areas sorediate. Fig. 365. .... *Cladonia bacillaris* (Ach.) Nyl.

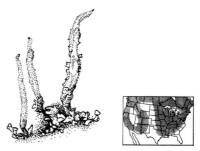

Figure 365

Figure 365   Cladonia bacillaris (×2)

Podetia whitish mineral gray to white, 0.5-1.5 cm tall, unbranched to sparingly branched toward the apex, a few squamules developed at the base; soredia farinose, diffuse; primary squamules poorly developed to dense, small; pycnidia common, apothecia rare. K+ yellowish, P— or P+ yellowish (barbatic acid, rarely also didymic and usnic acids). Very common on fenceposts, shingles, and the base of trees in open areas throughout North America. *C. macilenta* differs only in chemistry. Rare *C. pseudomacilenta* Asah., known from the northern Rockies, will also key here; it contains squamatic acid.

46a   (38) **Surface of podetia sorediate (use lens).** 47

46b   **Surface of podetia without soredia (or podetia finely squamulose or lacking).** ..
.... **56**

47a   (46) **Podetia generally unbranched, 1-2 cm tall or less.** 48

47b   **Podetia long and slender, usually branched several times (sometimes unbranched in *Cladonia cornuta*), 3-8 cm tall.** 53

48a   (47) **Podetia very short (1-3 mm), tipped with brown apothecia. Fig. 366.**
.... *Cladonia parasitica* (Hoffm.) Hoffm.

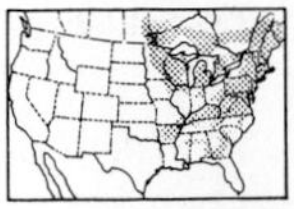

Figure 366

Figure 366   Cladonia parasitica

Podetia greenish mineral gray, covered with fine isidioid squamules; primary squamules finely incised, forming a dense mat, the margins becoming isidioid; apothecia common. K+, P+ deep yellow (thamnolic acid). Common on rotten logs and stumps in deep woods. A typical habitat is the cut surface of old stumps and it is difficult to cut across the grain to get good specimens. *C. caespiticia* differs in chemistry (K—) and less divided squamules. *C. botrytes* is also K— but has pale flesh colored apothecia.

48b   **Podetia more than 5 mm tall, apothecia rare.** **49**

49a   (48) **Primary squamules conspicuous (2-5 mm long), usually dark green, podetia arising from the centers. Fig. 367.** ....
...... *Cladonia coniocraea* (Flk.) Spreng.

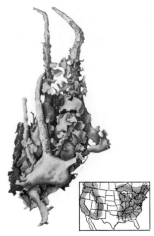

Figure 367

Figure 367　Cladonia coniocraea　(×1.5)

Podetia whitish green, 0.5-2.0 cm tall, pointed or forming narrow irregular cups, the soredia farinose, corticate areas restricted to the base; pycnidia common, apothecia rare, dark brown. P+ red (fumarprotocetraric acid). Common on humus and rotting logs in closed woods. This is probably the most frequently collected pointed *Cladonia* without apothecia. *C. ochrochlora* Flk. is virtually identical except for larger esorediate corticate areas, amounting to one-third or more of the podetial surface. *C. incrassata* is similar and may key here. Check carefully for red pycnidia. A rarer species in the western states, *C. bacilliformis* (Nyl.) Vain., has identical morphology but is yellow (usnic and barbatic acids).

49b　Primary squamules small, less than 2 mm long, sometimes finely divided, greenish mineral gray. 50

50a　(49) Podetia blunt, the base with coarse isidioid granules. Fig. 368. ....
...... *Cladonia cylindrica* (Evans) Evans

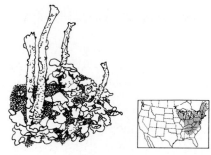

Figure 368

Figure 368　Cladonia cylindrica　(×1.5)

Podetia ashy white, 0.5-1.5 cm high, rarely branched, the tips blunt or slightly expanded into tiny shallow cups, diffusely sorediate, sparsely squamulate toward the base; primary squamules small, incised, usually well developed; pycnidia common, apothecia rare, dark brown. P+ red (fumarprotocetraric and grayanic acid). Very common on stumps, base of trees, rotten logs, and humus in open woods and pastures. Podetia are often only poorly developed. *C. coniocraea* is very similar but in general its squamules are much larger and greener with grayanic acid lacking. *C. balfourii* has more pointed podetia and also lacks grayanic acid.

50b　Podetia pointed or tipped with apothecia. 51

51a　(50) Soredia powdery, more or less continuous over the podetial surface; southern United States. Fig. 369. ....
.... *Cladonia balfourii* Cromb.

Figure 369

Figure 369   Cladonia balfourii (×1)

Podetia ashy white, 1-2 cm tall, rarely branched, pointed or forming tiny cups, more or less diffusely sorediate, sparsely squamulate toward the base; primary squamules small, incised, often well developed; pycnidia common, apothecia rather rare, dark brown. P+ red (fumarprotocetraric acid). Widespread on sandy soil or humus in open areas. This rather undistinguished species is very similar to *C. cylindrica*, which differs chiefly in producing grayanic acid as well as fumarprotocetraric. *C. pityrea* has ecorticate bare areas.

51b   Soredia more granular, scattered, bare ecorticate patches visible; eastern and northern North America. 52

**52a   (51) Podetia usually small, sometimes irregular and twisted, more or less free of squamules. Fig. 370. ....
.... *Cladonia pityrea* (Flk.) Fr.**

Figure 370

Figure 370   Cladonia pityrea (×1.5)

Podetia whitish mineral gray, 0.5-1.5 cm high, with diffuse coarse soredia and tiny squamules but extensive ecorticate areas developing; primary squamules sparse, tiny, incised; apothecia rare. P+ red (fumarprotocetraric acid). Widespread on soil, old logs, and the base of trees in open woods. There is a great range of variation in production of soredia and squamules. When poorly developed, this *Cladonia*

is hard to separate from *C. coniocraea* and other species with pointed podetia. A useful character to distinguish it would be the tinier incised squamules, almost identical with those of *C. squamosa*.

**52b   Podetia larger, more or less squamulose. Fig. 371. ....
...... *Cladonia decorticata* (Flk.) Spreng.**

Figure 371

Figure 371   Cladonia decorticata (×1.5)

Podetia whitish mineral gray, 1.5-3 cm high, sparingly branched in the upper parts, sorediate, the soredia granular and intergrading with areoles, the cortex fissured or with longitudinal hyphae apparent; primary squamules medium-sized; apothecia and pycnidia rare, dark brown. K−, P− (perlatolic acid). Widespread on soil or humus in conifer forests or open areas. Some forms of *C. rei* may key here; they would lack fissures on the podetia and would be squamulose mostly in the lower half. *C. cylindrica* is similar but lacks squamules on the podeta and is P+ red. *C. acuminata* (Ach.) Norrl. (K+ yellow→red, norstictic acid present) and *C. norrlinii* Vain. (K+ yellow, atranorin, and a P+ yellow unknown) are scarcely distinguishable rare chemical variants in the northern forests.

**53a   (47) Podetia yellowish. Fig. 372. ....
.... *Cladonia cyanipes* (Somm.) Nyl.**

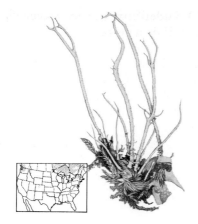

Figure 372

Figure 372   Cladonia cyanipes (×2)

Podetia 2-5 cm tall, sparingly branched and sorediate in the upper parts, squamulate toward the base; primary squamules medium-sized to small, not well developed; pycnidia common, apothecia very rare, dark brown. Widespread on humus, rotting logs, and soil in open areas and bogs. This species is very similar to *C. cornuta* in appearance but differs in chemistry.

53b   Podetia whitish to greenish mineral gray or brownish. **54**

54a   (53) Large areas corticate toward the base; soredia mostly toward tips. Fig. 373. .... ***Cladonia cornuta*** **(L.) Hoffm.**

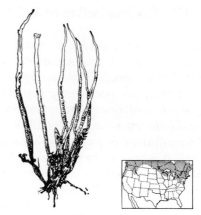

Figure 373

Figure 373   Cladonia cornuta (×1)

Podetia greenish to brownish mineral gray, 3-8 cm tall, sparsely squamulate, infrequently branched; primary squamules medium-sized, poorly developed or absent; pycnidia and apothecia rare, dark brown. P+ red (fumarprotocetraric acid). Widespread on humus or among mosses over rocks in open areas. The elongate thin podetia resemble forms of *C. gracilis*, which lacks soredia.

54b   Soredia occurring over most of the podetial surface. **55**

55a   (54) Podetia usually dichotomously branched. Fig. 374. .... **.. *Cladonia scabriuscula* (Duby) Leight.**

Figure 374

Figure 374 Cladonia scabriuscula (×1)

Podetia light mineral gray, 5-8 cm tall, sparsely to densely covered with squamules, moderately sorediate, especially in the upper parts; primary squamules poorly developed or absent; pycnidia common, apothecia very rare. P+ red (fumarprotocetraric acid). Widespread on soil in open pastures. Formerly confused with the more common *C. farinacea,* this species has fewer soredia and numerous squamules on the podetia. *C. subulata* (L.) Wigg., a rare northern species, differs chiefly in having almost entirely sorediate podetia.

55b Podetia irregularly branched to furcate. Fig. 375. .... *Cladonia farinacea* (**Vain.**) **Evans**

Figure 375

Figure 375 Cladonia farinacea (×1.5)

Podetia light mineral gray, 3-10 cm tall, sparingly branched with open axils; becoming sparsely squamulose toward the base; soredia farinose, diffuse over much of the surface; primary squamules small, dissected; apothecia rare, dark brown. P+ red (fumarprotocetraric acid). Common on soil and mossy rocks in pastures and along roadsides. *C. glauca* Flk., which occurs only as far south as Connecticut, is very similar except for chemistry (P− squamatic acid). *C. scabriuscula* has granular, sparser soredia and more numerous squamules but shares the same P+ red reaction.

56a (46) Podetia simple to branched, usually well developed. 57

56b Podetia barely developed, very short, or lacking and only large sterile squamules collected. 68

57a (56) Podetia more or less richly branched, apothecia inconspicuous. Fig. 376. .. *Cladonia furcata* (**Huds.**) **Schrad.**

Figure 376

Figure 376 Cladonia furcata (×1)

Podetia greenish to brownish mineral gray, sparsely squamulate, the axils open, sometimes expanding into narrow irregular cups, 4-8 cm tall; primary squamules sparsely developed to lacking; pycnidia common, apothecia rare, dark brown. P+ red (fumarprotocetraric acid). Very common on mossy rocks and humus in mature forests and fields or along roadsides. This will be collected very frequently and because of the great range of variability in size and in the development of squamules on the podetia some practice will be needed for positive identification. Two poorly known chemical variants will be collected in the northern part of the range: *C. pseudorangiformis* Asah. (atranorin, merochlorophaeic and psoromic acids) and *C. subrangiformis* Sandst. (atranorin and fumarprotocetraric acid).

57b Podetia usually simple or sparingly branched, often tipped with large apothecia. 58

58a (57) Surface of podetia finely and densely squamulose (use hand lens); UV+ white. .... (p. 188) *Cladonia squamosa*

58b Surface of podetia coarsely squamulose or free of squamules; UV−. 59

59a (58) Podetia and squamules with a yellowish cast (usnic acid present). Fig. 377. .... *Cladonia piedmontensis* Merr.

Figure 377

Figure 377 Cladonia piedmontensis (×1)

Podetia 1-3 cm tall, sparingly branched, moderately squamulate over most of the surface; primary squamules medium-sized, usually well developed; pycnidia and apothecia common, brown. Common on sandy soil in open woods and along roadbanks. No other Cladonias in eastern United States have consistently well developed yellow podetia capped with apothecia.

59b Podetia and squamules whitish to greenish mineral gray (usnic acid lacking except for *C. botrytes*). 60

60a (59) Primary squamules conspicuous, generally large and strap-shaped, 4-10 mm long; surface of podetia smooth. .. 61

60b Primary squamules smaller, about 1 mm long, rounded and crowded; surface of podetia finely areolate with longitudinal strands of cortex usually visible. 64

61a (60) Podetia lacerated and perforate; squamules very large, often grow free of the substrate. Fig. 378. .... *Cladonia turgida* (Ehrh.) Hoffm.

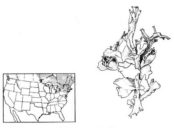

Figure 378

Figure 378 Cladonia turgida (×1)

Podetia aborted to well developed, loosely attached, to 5 cm tall, moderately branched with axils expanding into irregular cups, sparsely squamulate; primary squamules large, 1-3 cm long, dense, chalky white on the lower surface; pycnidia common, apothecia rare, dark brown. K+ yellow (atranorin), P− or P+ red (fumarprotocetraric acid). Widespread on humus or soil over rocks in open areas. The squamules are often so well developed as to be taller than the podetia.

61b Podetia entire, squamules smaller, usually firmly attached to soil. 62

62a (61) Podetia more or less free of squamules, usually tipped with apothecia. 63

62b Podetia becoming squamulose along the entire length; apothecia inconspicuous or lacking. 66

63a (62) Podetia and squamules (lower surface) K+ yellow or red. Fig. 379. .... *Cladonia polycarpoides* Nyl.

Figure 379

Figure 379   Cladonia polycarpoides

Podetia commonly not developed, simple to sparingly branched, 0.5-1.5 cm high; primary squamules well developed, greenish mineral gray, strap-shaped, up to 1 cm long, the lower surface white to buff, forming extensive mats; apothecia common. K+ yellow→red, P+ orange (norstictic acid). Common on soil in abandoned fields, open woods, and along roadsides. This is probably the most frequently collected sterile *Cladonia* on open soil. Two very close relatives in southeastern United States, *C. polycarpia* Merr. (atranorin, norstictic and stictic acids) and *C. clavulifera* (below), would have to be distinguished with microchemical tests. Another one, *C. symphycarpa* (Ach.) Fr. (K+, P+ yellow, psoromic acid and atranorin), occurs in Kansas, Colorado, and Michigan.

63b Podetia and squamules K−. Fig. 380. .... *Cladonia clavulifera* Vain.

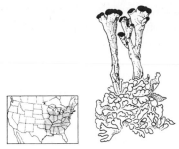

Figure 380

Figures 380   Cladonia clavulifera (×1.5)

Podetia greenish mineral gray, 1-2 cm tall, simple or sparingly branched toward the upper parts, sparsely squamulate; primary squamules well developed, medium-sized; apothecia very common, dark brown. P+ red (fumarprotocetraric acid). Widespread on soil along roadbanks or in open fields. Podetia are usually present but this common lichen may be collected only as extensive patches of squamules.

64a (61) **Apothecia dark brown to blackish; areoles strongly contrasting with lighter colored cortex. Fig. 381.** .... *Cladonia cariosa* (Ach.) Spreng.

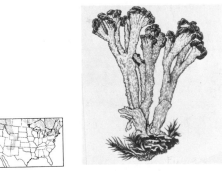

Figure 381

Figure 381   Cladonia cariosa (×2)

Podetia whitish mineral gray, 1.5-2.5 cm tall, the surface with scattered areoles partially covering the ribbed medulla; primary squamules small, sparsely developed; apothecia conspicuous, dark brown. K+, P+ yellowish (atranorin). Common on soil in open fields and along road cuts. While sometimes confused with *C. capitata, C. cariosa* has much darker apothecia and the podetia are not noticeably twisted.

64b Apothecia pale, tan to flesh-colored; areoles not standing out conspicuously. ..
.... **65**

65a (64) Podetia 1-1.5 cm tall; collected in eastern North America. Fig. 382. ....
...... *Cladonia capitata* (**Michx.**) **Spreng.**

Figure 382

Figure 382   Cladonia capitata (×2)

Podetia mineral gray, 1.0-1.5 cm high, sparingly branched, with small dispersed corticate areas partially covering twisted, cartilaginous ribs; primary squamules very small, poorly developed; apothecia conspicuous. P+ red (fumarprotocetraric acid). Common on soil or at the base of trees in mature forests. The pale

brown apothecia are diagnostic for this common *Cladonia*. *C. cariosa,* a northern species, has similar capitate apothecia but they are dark brown. Another comparable species, *C. brevis* (Sandst.) Sandst., has shorter podetia and a different chemistry (K−, P+ yellow, psoromic acid).

65b Podetia very short, about 5 mm high; collected in boreal eastern and western North America. Fig. 383. ....
.... *Cladonia botrytes* (**Hag.**) **Willd.**

Figure 383

Figure 383   Cladonia botrytes (×1.5)

Podetia light mineral to yellowish gray, 0.3-0.6 cm tall, sparingly branched in the upper part; primary squamules scattered, tiny; apothecia common, tan or flesh-colored. K−, P− (usnic and barbatic acids). Widespread, but never abundant, on rotten logs and stumps in open woods.

66a (62) Collected in alpine-arctic localities. Fig. 384. ....
Cladonia macrophylla (**Schaer.**) **Stenh.**

Figure 384

Figure 384   Cladonia macrophylla (×1.5)

Podetia whitish mineral gray, 2-4 cm high, sparingly branched in the upper parts, densely areolate-squamulate but conspicuous ridged cartilaginous areas visible; primary squamules medium-sized, scattered; pycnidia common, apothecia rare, dark brown. P+ yellow (psoromic acid). Widespread on soil and humus in exposed areas. The rather irregular often gnarled podetia are characteristic of the species. *Cladonia squamosa* can be separated by a P— reaction and lack of medulla ridges.

66b   Collected in eastern United States. 67

67a   (66)   Primary squamules well developed; podetia sparingly branched. Fig. 385. .... *Cladonia beaumontii* Tuck.

Figure 385

Figure 385   Cladonia beaumontii

Podetia whitish mineral gray, 0.5-1.5 cm tall, covered with tiny squamules; primary squamules well developed, medium-sized, incised; pycnidia and apothecia rare, brown. P+ yellow (baeomycic and squamatic acids). On sandy soil, rotten logs, and base of trees in open areas.

A rare companion species is *C. santensis* Tuck., which has smaller podetia (up to 0.8 cm) and reacts K+ yellow (thamnolic acid).

67b   Primary squamules lacking; podetia becoming moderately branched. .... *Cladonia squamosa* (see p. 183)

68a   (56)   Apothecia short-stalked (1-2 mm high) or sessile on the squamules (under hand lens). 69

68b   Podetia and apothecia lacking, only large squamules present. .... (Key V, p. 230) *Squamulose Lichens*

69a   (68)   Squamules finely divided and incised. Fig. 386. .... *Cladonia caespiticia* (Pers.) Flk.

Figure 386

Figure 386   Cladonia caespiticia

Podetia very tiny, up to 0.1 cm tall; primary squamules 0.3-0.5 cm long, incised, forming a dense mat; apothecia common, dark brown. P+ red (fumarprotocetraric acid). Common on rotting logs and on mosses over rocks in shady woods. When sterile, the squamules can be confused with either *C. parasitica* (K+ deep yellow) or aborted *C. squamosa* (K−, P−).

69b Margins of squamules entire, not finely divided. **70**

70a (69) Squamules with a yellowish cast above and on the lower surface (usnic acid present); K−, P−, Fig. 387. .... *Cladonia robbinsii* Evans

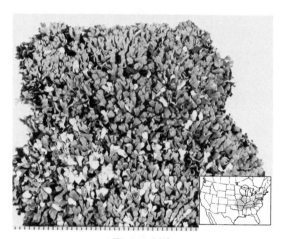

Figure 387

Figure 387   Cladonia robbinsii

Podetia rarely present, about 1 cm tall, irregularly cup-shaped; primary squamules well developed, strap-shaped, about 0.5 cm long, forming mats 3-6 cm broad, the lower surface cream to faint yellowish buff; apothecia rare, dark brown. Widespread on open soil or over rocks. The greenish yellow color will distinguish this rather rare lichen from other sterile *Cladonia* mats except for *C. strepsilis,* which

forms more compact colonies and reacts C+ green.

70b Squamules greenish mineral gray to brownish, lower surface chalky white to brownish. **71**

71a (70) Lower surface of squamules K+ yellow or yellow→red. **72**

71b Lower surface of squamules K− or K+ brownish. **74**

72a (71) Squamules very long (2-3 cm) and narrow with a blackened base. .... (p. 231) *Gymnoderma lineare*

72b Squamules shorter, broad, up to 1 cm long, the base not blackened. **73**

73a (72) Lower surface of squamules K+ yellow→red. .... (p. 193) *Cladonia polycarpoides*

73b Squamules persistently K+ yellow. .... (p. 192) *Cladonia turgida*

74a (71) Lower surface of squamules chalky white, C−; colonies scattered. Fig. 388. .. .. *Cladonia apodocarpa* Robb.

Figure 388

Figure 388   Cladonia apodocarpa (×1)

Figure 389   Cladonia strepsilis

Podetia lacking; primary squamules greenish mineral gray, large and strap-shaped, about 1 cm long, suberect with the white lower surface visible; apothecia rare, borne on tiny stipes on the squamules. P+ red (fumarprotocetraric acid). Very common on sandy soil in open woods or along roadsides. This is usually collected as sterile squamules. Closely related sterile species such as *C. polycarpoides*, *C. clavulifera*, *C. mateocyatha*, and *C. subapodocarpa* Harris, which occurs on rocks in the Appalachians and contains perlatolic acid (K−, P−), will have to be separated by appropriate chemical tests. In the western states, *Psora novomexicana* (see page 235) has similar but more rounded squamules and large adnate apothecia.

74b   Lower surface cream to brownish, C+, KC+ green; colonies often forming compact balls or mats. Fig. 389. ....
.... *Cladonia strepsilis* (Ach.) Vain.

Podetia rarely developed, irregularly cup-shaped, up to 1 cm tall, primary squamules well developed, yellowish to greenish gray, the lower surface whitish to cream, often forming compact mats or balls 4-6 cm broad; apothecia very rare, dark brown. C+ green (strepsilin), P+ yellow (baeomycic acid). Widespread on soil in open areas and between boulders at large outcrops. The most unusual feature is the C+ green reaction, but the growth habit alone is diagnostic. *C. robbinsii* has a similar color but differs chemically. *C. apodocarpa* has longer squamules chalky white below.

75a   (17) Thallus chartreuse to sulphur yellow or yellowish green (becoming reddish tinged in some *Usnea* species) (usnic acid present in all groups except *Letharia*). 76

75b   Thallus white, or whitish to brownish gray (usnic acid lacking). 126

76a   (75) Thallus chartreuse to deep sulphur yellow (see Frontispiece No. 4); collected only in western North America. ....
.... 77

76b   Thallus pale to greenish yellow or yellowish green; collected throughout North America. 78

77a   (76) Thallus diffusely sorediate; apothecia rare. Fig. 390. ....
.... *Letharia vulpina* (L.) Hue

Figure 389

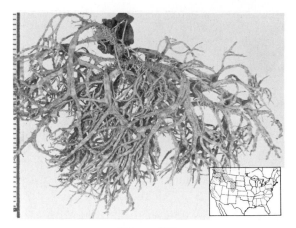

Figure 390

Figure 390 Letharia vulpina

Thallus tufted, to 10 cm broad but often covering very large areas of trunk and branches; branches becoming wrinkled; medulla white. Cortex and medulla, K–, C–, P– (vulpinic acid). Very common on conifers, especially on lower trunks and dead branches. The nonsorediate relative *L. columbiana* (below) usually grows higher up on the trunks. These two species together form spectacular displays in the California redwood and *Sequoia* forests.

**77b Thallus lacking soredia; apothecia large and abundant. Frontispiece No. 4. ....
.... *Letharia columbiana* (Nutt.) Thoms.**

Thallus chartreuse to greenish lemon yellow, usually erect and tufted on branches, to 10 cm broad but often covering large areas; branches dull, irregularly flattened, rugose; medulla white; apothecia invariably present, the rim lobulate. Cortex and medulla K–, C–, P– (vulpinic acid). Common and conspicuous on exposed conifers and dead branches, usually out of reach in the canopy in dense forests.

**78a (76) Thallus growing more or less free on soil or humus or over humus on rocks.
.... 79**

78b Thallus basally attached to trees, rocks, and rarely soil. 92

**79a (78) Thallus flattened and strap-shaped. Fig. 391. ....
.... *Cetraria cucullata* (Bell.) Ach.**

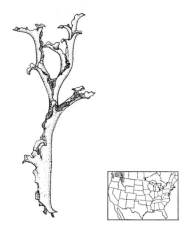

Figure 391

Figure 391 Cetraria cucullata (×1.5)

Thallus bright greenish yellow, 3-5 cm tall, growing erect among mosses or on humus; surface shiny, smooth to faintly rugose and pitted; base sometimes turning dark purple red; pycnidia very common, tiny; apothecia rare. Cortex and medulla K–, C–, P– (usnic acid only). Widespread in arctic and arctic-alpine regions. Closely related *C. nivalis* (L.) Ach., which has more flattened and deeply rugose lobes, frequently occurs with this species. *C. tilesii* Ach. differs in having a deep lemon or orange yellow thallus (vulpinic acid); it will be found in the high Rockies.

79b Thallus round in cross section (examine with hand lens). 80

80a (79) Surface of branches dull and fibrous, lacking a continuous cortex (use hand lens and compare with Fig. 392). 81

Figure 392

Figure 392   Fibrous surface of *Cladina rangiferina* (left) and corticate surface of *Cladonia dimorphoclada* (×10)

80b Surface of branches shiny, corticate, continuous or areolate (may be covered with fine hairs in *Cladonia boryi* and *C. pachycladodes*). 84

81a (80) Thallus forming compact heads 2-3 cm in diameter without conspicuous main branches. Fig. 393. ....
.... *Cladina stellaris* (Opiz) Brodo

Figure 393

Figure 393   Cladina stellaris (×½)

Podetia pale yellowish gray, 6-10 cm high, the colonies separate or loosely clumped; pycnidia common, apothecia very rare. On soil and mossy humus in open woods or pastures. This unmistakable species is rare in the southern part of its range. From North Carolina to Texas a very similar species, *C. evansii*, grows on sterile white sand; it is ashy white and lacks usnic acid but contains atranorin and perlatolic acid. Both of these species are frequently collected for use as imitation shrubbery and trees in model train layouts. This species was previously called *Cladonia alpestris*.

81b Thallus forming extensive entangled colonies without discrete heads but with distinct main branches. 82

82a (81) Axils of branches closed (use hand lens); ultimate branches mostly in pairs. Fig. 394. ....
Cladina subtenuis (Abb.) Hale & Culb.

Figure 394

Figure 394 Cladina subtenuis

Podetia 4-8 cm tall, often forming mats up to a foot in diameter; pycnidia common, apothecia very rare. P+ red (fumarprotocetraric acid). Common on sandy soil in open pine forests and along roadsides. This is undoubtedly the most frequently collected "Reindeer Moss" in southeastern United States, often growing with *C. rangiferina*. In mountainous areas and the northern part of its range it may occur with *C. arbuscula* or *C. mitis*, from which it must be carefully distinguished by branching pattern, axils, and chemical tests. In the Pacific Northwest one may collect *C. tenuis* (Flk.) Hale & Culb., a very closely related species with more distinct main stems.

82b   Axils of branches open (use hand lens); ultimate branches grouped in 3's and 4's. 83

83a   (82) Ultimate branches rather coarse, strongly pointing in one direction. Fig. 395. .... *Cladina arbuscula* (Wallr.) Hale & Culb.

Figure 395

Figure 395 Cladina arbuscula (×1)

Podetia 6-10 cm tall, the colonies often extensive, the ultimate branches mostly pointing in one direction; pycnidia common, apothecia very rare, dark brown. P+ red (fumarprotocetraric acid). Common on soil and humus in open pastures and fields. Without a P test, this species is difficult to separate from *C. mitis*, which generally has ultimate branches pointing in various directions.

83b   Ultimate branches rather fine, not strongly pointing in one direction. Fig. 396. .... *Cladina mitis* (Sandst.) Hale & Culb.

Figure 396

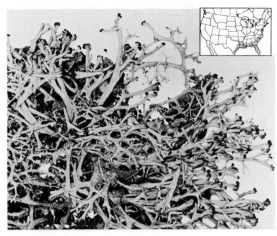

Figure 397

Figure 396   Cladina mitis (×1)

Podetia 4-8 cm tall, forming extensive colonies; branches generally in 3's, rarely 4's. Cortex K−, P− (usnic and fatty acids). Widespread in open pine forests and open fields. This is the most commonly collected *Cladina* in the western states, being supplanted over most of eastern United States by *C. subtenuis*. If paraphenylenediamine is available, one can quickly identify this species by the negative reaction. Both *C. arbuscula* and *C. subtenuis* are P+ red. There are several other P− species, all mostly arctic or boreal but extending into New England and the Great Lakes region: *C. impexa* (Harm.) Lesd., which has perlatolic acid (TLC test needed), *C. submitis* (Evans) Hale & Culb., which is a much coarser plant with branches in 4's, and *C. terrae-novae* (Ahti) Hale & Culb., which reacts K+ yellow (atranorin present). Along the coast from Oregon into British Columbia one will collect the conspicuous *C. pacifica* (Ahti) Hale & Culb., a robust yellowish species with perlatolic acid.

84a   (80) Apothecia and/or pycnidia present, red (use hand lens), reacting K+ purple. Fig. 397. .... *Cladonia leporina* Fr.

Figure 397   Cladonia leporina (×1)

Podetia yellowish green, leathery, prostrate, 3-6 cm high, forming mats 5-10 cm broad; pycnidia and apothecia common, dark red. P+ yellow (baeomycic acid). Common on sandy soil and stumps in open areas. This is frequently collected and can be recognized at sight. *C. caroliniana* has a similar aspect but differs in being very brittle and lacking red apothecia.

84b   Apothecia and pycnidia brown, K−, or lacking. 85

85a   (84) Branches solid (section with razor blade); collected in the western Great Plains and northern Rockies. Fig. 398. .... *Agrestia hispida* (Meresch.) Hale & Culb.

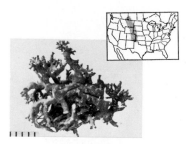

Figure 398

Figure 398   Agrestia hispida

Thallus dark greenish or olive yellow, prostrate, brittle, 1-2 cm broad; branches spinulate, irregularly thickened, with white pores; apothecia rare, sessile to immersed. Widespread in the Great Plains. Long overlooked, this unusual lichen is apparently a common component of the prairie soil vegetation as well as semi-desert areas in the Rocky Mountains.

85b   Branches hollow; collected throughout North America. **86**

86a   (85) Thallus thick and inflated, finger-like and unbranched, faint brownish yellow. Fig. 399. ....
.... ***Dactylina arctica*** (Hook.) Nyl.

Figure 399

Figure 399   Dactylina arctica (×1)

Thallus 2-3 cm tall, scattered or growing in small mats 4-8 cm broad; surface of branches smooth and shiny; apothecia very rare. Cortex and medulla P+ red (fumarprotocetraric acid) or P−, K−, C+ rose (gyrophoric acid). This is

one of the best known arctic lichens, occurring in arctic Canada and southward at higher elevations in Wyoming. *Dactylina madreporiformis* (Ach.) Tuck. is a smaller plant with blunt tips, common in the alpine zone in the Rockies. A third species, *D. ramulosa* (Hook.) Tuck., is white-pruinose and occurs in the arctic and at high elevations in the northern Rockies.

86b   Thallus usually thin and branched, 0.5-1 mm thick (inflated only in *Cladonia boryi* and *C. caroliniana*), pale yellowish green. **87**

87a   (86) Tips of podetia flaring into shallow cups. Fig. 400. ....
.... ***Cladonia amaurocraea*** (Flk.) Schaer.

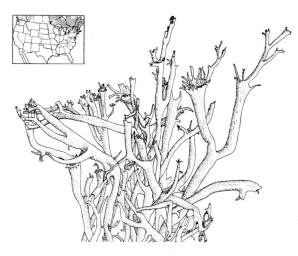

Figure 400

Figure 400   Cladonia amaurocraea (×1.5)

Podetia 6-10 cm tall, erect, very brittle when dry, completely corticate and shiny, sparingly branched, the axils open, expanding into proliferating cups; pycnidia common but apothecia rare. Cortex K−, C−, P− (usnic and barbatic acids). On soil and humus over rocks in open areas. In the southern part of its range, this species becomes quite rare but can be confused with *C. uncialis,* which lacks true cups and has shorter blunter tips.

87b **Tips of podetia pointed, not cup-forming.** **88**

88a **(87) Podetia coarse and inflated, 2-4 mm in diameter, irregularly perforated.** **89**

88b **Podetia finer, 0.5-1 mm in diameter, without perforations.** **90**

89a **(88) Fine hairs present on surface (use hand lens and examine tips). Fig. 401.** .... *Cladonia boryi* **Tuck.**

Figure 401   Cladonia boryi (×3)

Podetia light yellowish green, 4-8 cm tall, sparingly branched, the surface pitted and in part ecorticate; pycnidia common but apothecia rare. Cortex K−, C−, P− (usnic acid). On sandy soil in openings in the forest. This rather rare species can be recognized by the unusually thick podetia and the pitted surface.

89b **Surface hairs lacking. Fig. 402.** .... *Cladonia caroliniana* **(Schwein.) Tuck.**

Figure 402

Figure 402   Cladonia caroliniana (×1)

Podetia greenish yellow, brittle, erect to prostrate, 3-6 cm high, irregularly inflated, and lacerated, smooth or with bristle-like secondary branches; pycnidia common, apothecia rare, dark brown. K−, P− (usnic and squamatic acids). Common on sandy soil and humus in open areas. In the typical irregularly inflated form this species is unmistakable, growing in large mats along the edges of large rock outcrops, but forms with narrow podetia intergrade with *C. uncialis* and *C. dimorphoclada* to such an extent that the species can hardly be distinguished.

90a **(88) Surface of podetia dull, covered with fine hairs (use hand lens). Fig. 403.** .... *Cladonia pachycladodes* **Vain.**

Figure 403

Figure 403   Cladonia pachycladodes (×2)

Podetia pale greenish yellow, prostrate, 1-2 cm tall, forming mats 5-10 cm broad; cortex covered with a fine tomentum composed of loose hyphae; pycnidia and apothecia very rare. Cortex K−, C−, P− (usnic acid). Widespread on barren sandy soil in the Coastal Plain. It occurs intermingled with *C. leporina* in many areas, and can be distinguished by the more whitish cast, brown pycnidia (when present), and fine hairs.

90b   Surface of podetia shiny, without hairs. ..
.... **91**

91a   (90) Podetia very fine, almost hairlike, 0.3-0.6 mm in diameter. Fig. 404. ....
.... *Cladonia subsetacea* **Evans**

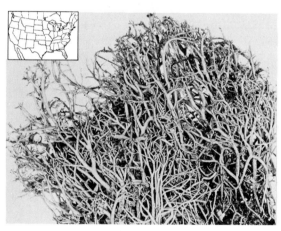

Figure 404

Figure 404   Cladonia subsetacea (×1)

Podetia light greenish yellow, prostrate, 1-2 cm tall, forming mats 4-8 cm broad; pycnidia and apothecia very rare. P+ yellow (in Florida specimens P−) (squamatic acid with or without baeomycic acid). Widespread on open sandy soil in the Coastal Plain from North Carolina to Florida. Hair-like fine podetial branches characterize this rather rare *Cladonia*.

91b   Podetia thicker, 1-1.5 mm in diameter. Fig. 405. ....
.... *Cladonia uncialis* **(L.) Wigg.**

Figure 405

Figure 405   Cladonia uncialis (×1)

Thallus yellowish green, usually more or less prostrate and forming mats 20 cm or more across, very brittle, richly branched; surface of podetia shiny; pycnidia common, apothecia rare. Cortex K−, C−, P− (usnic and squamatic acids; white fluorescence produced under a UV lamp). Very common on soil and among mosses in open areas. Rather robust forms of this species, previously identified as *C. caroliniana*, lack squamatic acid and have longitudinal ribs of medullary tissue on the inner surface of the podetia; these can be called *C. dimorphoclada* Robbins. Typical *C. uncialis* has a smooth inner surface, but this character is not always convincing and many intermediates will be found.

92a (78) Branches more or less strongly flattened or irregular in cross section (lens usually not needed), both sides the same color or one side paler or darker. 93

92b Branches round in cross section, without an upper and lower surface. 107

93a (92) Branches with marginal or laminal soredia. 94

93b Branches smooth, without soredia. 99

94a (93) Soredia granular to subsidiate, diffuse; branches irregularly rounded, limp. Fig. 406. .... *Evernia mesomorpha* Nyl.

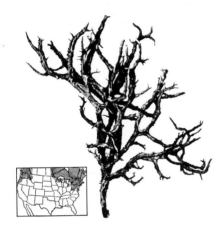

Figure 406

Figure 406   Evernia mesomorpha (×1)

Thallus pale yellowish green, erect to pendulous, 4-6 cm long; surface irregularly wrinkled; apothecia very rare. Cortex and medulla K−, C−, P− (usnic and divaricatic acids). Common on conifers, hardwoods, and fenceposts in the northern forests. Lack of fibrils and the irregularly thickened branches separate it from *Usnea*. A close relative, *E. divaricata* (L.) Ach., is a more pendulous, sparsely branched lichen rarely found at higher elevations in the Rocky Mountains.

94b Soredia in distinct marginal or apical soralia; branches flattened, leathery or brittle. 95

95a (94) Branches hollow and perforated (examine the basal area with a hand lens). Fig. 407. .... *Fistulariella roesleri* (Hochst.) Bowler & Rund.

Figure 407

Figure 407   Fistulariella roesleri (×3)

Thallus pale yellowish green, tufted, 1-4 cm long; branches attenuated at the tips with isidiate soredia; apothecia lacking. Cortex K−, C−, P− (divaricatic and usnic acids). Widespread on branches of conifers in open forests but not abundant. The hollow perforate branches will be overlooked at first in this inconspicuous little lichen.

95b Branches solid and imperforate. 96

96a (95) Tips of lobes bursting open, sorediate. Fig. 408. ....
.... *Ramalina pollinaria* (Westr.) Ach.

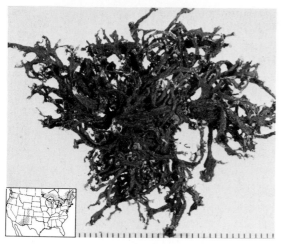

Figure 408

Figure 409

Figure 408   Ramalina pollinaria

Thallus pale yellowish green, tufted, 1-2 cm long; soredia diffuse, forming over the exposed medulla; apothecia very rare. Cortex K−, C− P− (usnic and evernic acids). Rare on tree trunks and sandstone outcrops in fairly sheltered areas. The species is easily recognized by the broad white sorediate patches. In California there is a very similar but larger species, *R. evernioides* Nyl., which contains a fatty acid (bourgeanic acid). This group, however, is not well known and undoubtedly more species will be discovered.

96b   Tips of lobes not bursting open; soralia mostly marginal. 97

97a   (96) Thallus growing on rocks. Fig. 409. .... *Ramalina intermedia* Nyl.

Figure 409   Ramalina intermedia

Thallus yellowish green, tufted, 1-3 cm long, often covering extensive patches on the overhanging part of ledges; branches becoming finely divided; apothecia rare. Cortex K−, C−, P− (sekikaic acid and usnic acid). Common on boulders and ledges in exposed areas, rarely at the base of trees. A chemical variant with protocetraric acid (P+ red), *R. petrina* Bowler & Rund., occurs in the southern Appalachians with the typical population.

97b   Growing on trees. 98

98a   (97) Thallus rather flabby; surface rugose and pitted. Fig. 410. .... *Evernia prunastri* (L.) Ach.

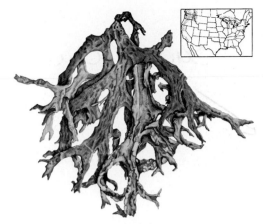

Figure 410

Figure 410 Evernia prunastri (×1)

Thallus pale yellowish green to gray, tufted, 2-7 cm long; branches 2-4 mm wide, the lower surface paler than the upper; apothecia rare. Cortex and medulla K−, C−, P− (evernic acid, usnic acid, traces of atranorin). Common on trunks and branches of trees and on fenceposts in open areas or in bogs. Though very common from California to Washington, this is a rarity in eastern North America, where *E. mesomorpha* is far more common.

98b Thallus more rigid; surface shiny, smooth or longitudinally striate. Fig. 411. .... *Ramalina farinacea* (L.) Ach.

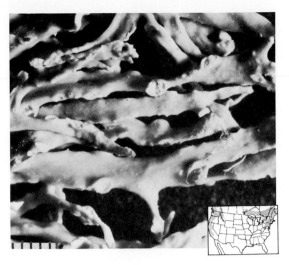

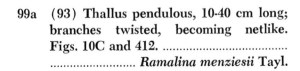

Figure 411

Figure 411 Ramalina farinacea

Thallus pale yellowish green, tufted to pendulous, 3-7 cm long; branches sparsely divided except toward the tips; apothecia very rare. Medulla K− or K+ yellow, P− or P+ red (with or without salazinic or protocetraric acids). Widespread on tree trunks, rarely on rocks, in open areas. This variable species is poorly known and appears to have several chemical races, including, for example, *R. hypoprotocetrarica* Culb. with hypoprotocetraric acid (K−, C−, P−). In California and Oregon into Washington *R. subleptocarpha* Rund. & Bowler will be found. It is a larger, much coarser plant reacting K−, P− (usnic acid and zeorin).

99a (93) Thallus pendulous, 10-40 cm long; branches twisted, becoming netlike. Figs. 10C and 412. .... *Ramalina menziesii* Tayl.

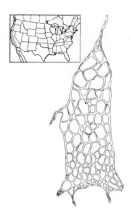

Figure 412

Figure 412   Ramalina menziesii (×1)

Thallus greenish yellow, often covering entire trees; branches soft, variable in width, becoming expanded and perforated to form a delicate network; apothecia rare. Cortex and medulla K−, C−, P− (usnic acid). Very common on oak trees and coastal conifers from California to Washington. This is one of the most spectacular lichens in the West. A similar pendulous species without perforations but with leathery strap-shaped branches, *R. usnea* (L.) Howe, will be collected in southern Florida, Texas, California, and Mexico.

99b   Thallus tufted, less than 10 cm long, not twisted or netlike. **100**

100a (99)   Branches   leathery,   sparsely branched; surface pitted and rugose; often collected on rocks. **101**

100b Branches thinner, moderately branched; surface smooth, papillate or striate but not rugose; collected on trees. **102**

101a (100) Branches strongly flattened with transverse cracks. Fig. 413. ....
Niebla homalea (Ach.) Rund. & Bowler

Figure 413

Figure 413   Niebla homalea (×1)

Thallus greenish yellow, turning brownish in the herbarium, blackening at the base, tufted, leathery, 4-10 cm long; surface becoming ridged and cracked; apothecia common, lateral. Medulla K−, C−, P− (barbatic and usnic acids). Common on rocks, especially along the seashore. It is often collected with *N. combeoides*, which is usually smaller and with rounded branches.

101b Branches   flattened   to   irregularly rounded, without transverse cracks. Fig. 414. .... *Niebla combeoides* (Nyl.) Rund. & Bowler

Figure 414

Figure 414   Niebla combeoides (×1)

Thallus greenish yellow, tufted, rather fragile, simple to sparingly branched, 2-4 cm long; surface of branches rugose; apothecia common, terminal. Medulla K+ yellow, P+ orange (stictic acid). Common in exposed habitats along the coast. A frequent companion species, *N. ceruchis* (Ach.) Rund. & Bowler, is larger (4-6 cm long), occurs on shrubs as well as rocks, and has mostly lateral apothecia. It contains a fatty acid (K−, P−) and produces a white cottony excrescence with age that resembles white mold. There are several other species in this unique genus, such as *N. cephalota* (Tuck.) Rund. & Bowler (with soredia), which are best developed in Baja California.

102a (100) **Branches generally broad and strap-shaped, 2-10 mm wide (use ruler).** **103**

102b **Branches narrow, 1.5 mm wide or less.** **106**

103a (102) **Branches with mostly laminal and marginal apothecia.** **104**

103b **Apothecia mostly terminal.** **105**

104a (103) **Collected in California. Fig. 415.** .... ***Ramalina leptocarpha* Tuck.**

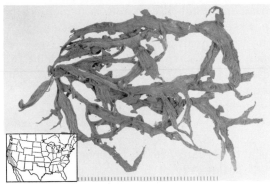

Figure 415

Figure 415   Ramalina leptocarpha

Thallus pale yellowish green, more or less pendulous, 5-10 cm long, the branches broad, 2-5 mm wide; surface rugose pitted, with some longitudinal striae; apothecia very common, mostly lateral. Cortex and medulla K−, P− (usnic acid only). On branches of oak trees and shrubs from San Francisco into Baja California.

104b **Collected in Texas. Fig. 416.** .... ***Ramalina ecklonii* (Spreng.) Mey. & Flot.**

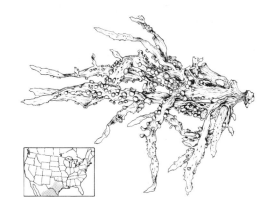

Figure 416

Figure 416   Ramalina ecklonii (×1)

Thallus greenish yellow, thin, membranous and rather soft, 4-8 cm long; surface deeply striate longitudinally; apothecia numerous, the spores ellipsoid. Cortex and medulla K−, P− (usnic acid only). Common on tree trunks in open areas. This species is common in the scrub forests of central and southern Texas. The laminal apothecia separate it from the *R. americana* group.

105a (103) Surface of branches smooth to striate. Fig. 417. .... *Ramalina americana* **Hale**

Figure 417

Figure 417   Ramalina americana (×1)

Thallus greenish yellow, tufted, 1-4 cm long; branches quite variable in width, 2-10 mm wide; surface striate, rarely perforate; apothecia common. Cortex and medulla K−, P− (usnic acid only). Common on tree trunks in open areas and on canopy branches in closed forests. This species was earlier called *R. fastigiata* and represents a large population that is still imperfectly known. Lobe width may vary up to 10 mm, with broad-lobed specimens often called *R. sinensis* Jatta. Striate pseudocyphellae (elongate white marks in the cortex) may be nearly absent to abundant. The typical species contains only usnic acid but other populations in southeastern United States may contain sekikaic acid or any one of five or six unidentified compounds, most K−, P−. A very similar species in Florida, *R. paludosa* Moore, contains paludosic acid.

105b Surface of branches strongly papillate or papillate-striate. Fig. 418. .... *Ramalina complanata* **(Sw.) Ach.**

Figure 418

Figure 418   Ramalina complanata (×1.5)

Thallus pale greenish yellow, tufted, leathery, 2-3 cm long; apothecia numerous, terminal. Cortex K−; medulla K−, P− (usnic and divaricatic acids). Common on tree trunks and branches in open areas and on palm trees along roadsides in Florida. Another southern species, *R. denticulata* Nyl., is also strongly papillate but generally larger (3-5 cm long) with narrower lobes; it contains salazinic acid (K+ yellow turning red, P+ orange). As with the *R. americana* group, however, the taxonomy of these species is still unsettled.

106a (102) Thallus 1-3 cm long; branches in part inflated and hollow. (use hand lens). Fig. 419. .... *Fistulariella* ...... *dilacerata* **(Hoffm.) Bowler & Rund.**

Figure 419

Figure 419   Fistulariella dilacerata (×1)

Thallus greenish yellow, tufted, the branches irregularly flattened to round with perforations toward the base; apothecia common. Cortex and medulla K—, C—, P— (usnic and divaricatic acids). Common on trees in open forests, especially in the northern Rockies. This is probably the most commonly collected *Fistulariella* in the Rocky Mountains. This genus was segregated from *Ramalina* on the basis of the hollow branches. There are several other species, especially in California, including *F. inflata* (Hook. & Tayl.) Bowler & Rund., that remain to be studied.

106b Thallus 2-6 cm long; branches solid. Fig. 420. .... *Ramalina willeyi* Howe

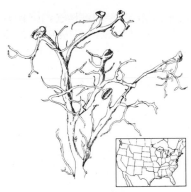

Figure 420

Figure 420   Ramalina willeyi (×1.5)

Thallus greenish yellow, tufted; branches narrow, becoming rounded, finely divided, the surface longitudinally striate; apothecia common, mostly subterminal or lateral, spores ellipsoid. Medulla K+ red, P+ orange (salazinic acid or if K— protocetraric acid). Common on shrubs and tree branches in the Coastal Plain. There are three closely related species that have nearly the same range: *R. tenuis* (Tuck.) Merr. with thinner attenuated branches; *R. montagnei* De Not. with fusiform spores; and *R. stenospora* Müll. Arg. with strongly flattened branches and fusiform spores.

107a (92) Thallus pendulous, 10-50 cm long, growing draped on tree branches or with some basal attachment. **108**

107b Thallus distinctly tufted, 3-10 cm long, firmly attached at the base. **113**

108a (107) Branches smooth, sparsely to richly branched but lacking fibrils. **109**

108b Branches with numerous short lateral fibrils and branchlets (see Fig. 484). ......
.... **111**

**109a (108) Branches loosely filled with me-
dulla, a central cord lacking (use hand
lens and section as shown in Fig. 427).
Fig. 421.** ....
.... *Alectoria sarmentosa* (Ach.) Ach.

Figure 421

Figure 421   Alectoria sarmentosa

Thallus greenish yellow, pendulous and free
growing but sometimes basally attached, 10-
30 cm long; branches white striate, often becoming twisted; apothecia not common. Cortex
and medulla K−, C−, P− (usnic and alectoronic or barbatic acids). Common on trees in
boreal and mountainous areas, often a good indicator of the level of the snow pack. Beginners
will invariably call this an *Usnea* but there is
no central cord. In the southern Appalachians
it is replaced by *A. fallacina* Mot., a very similar species distinguished by knobbier branches
and a dense medulla. In the Cascades *A. sarmentosa* grows with *A. lata* (Tayl.) Linds. and
A. vancouverensis Brodo and Hawks. The for-
mer is more prostrate and the latter C+ red
(olivetoric acid) but the Brodo and Hawksworth monograph should be consulted to separate them. In the arctic zone in Canada one
will collect *A. ochroleuca* (Hoffm.) Mass.,
a more greenish tundra species, and *A. nigri-*

cans (Ach.) Nyl., which is greenish black at
the tips but yellowish toward the base.

**109b Branches solid with a distinct, dense cen-
tral cord (use razor blade and examine
with hand lens).** **110**

**110a (109) Base of main branches smooth.
Fig. 422.** .... *Usnea trichodea* Ach.

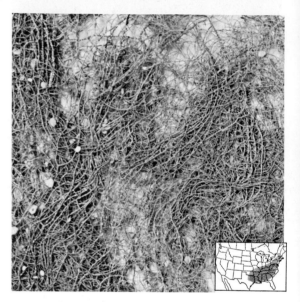

Figure 422

Figure 422   Usnea trichodea (×1)

Thallus pale greenish yellow, pendulous, soft,
10-30 cm long; main branches annulate,
cracked; apothecia common. Medulla K+ red
or K−, P+ orange or P− (usnic acid and
salazinic acid or unknowns). Widespread on
trees in the open, often in swamps or bayous.
This species is best developed in the southern
states where it may be confused with Spanish
Moss.

110b Base of main branches becoming rugose and foveolate. Fig. 423. ....
.... *Usnea cavernosa* **Tuck.**

Figure 423

Figure 423   Usnea cavernosa

Thallus pale greenish yellow, pendulous, soft, 10-40 cm long; branches continuous; apothecia occasional. Medulla K+ red, P+ orange or re-acting negative (usnic acid and salazinic acid or lacking substances). This is the most commonly collected pendulous *Usnea* in the northern states and Canada, growing on conifers.

111a (108) Main branches with sharp angular ridges. Fig. 424. .... *Usnea angulata* **Ach.**

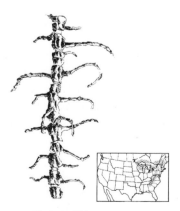

Figure 424

Figure 424   Usnea angulata (×3)

Thallus dark greenish yellow pendulous, stiff, 15-25 cm long; main branch distinct, with short secondary branchlets, the branchlets papillate and sparsely sorediate; apothecia rare. Medulla K+ yellow, P+ yellowish (norstictic acid). Rare in tree tops. This *Usnea* is easy to identify but it is not only rare but difficult to collect in tall trees. In Florida there is a pendulous species with a smooth cortex and articulated branches called *U. dimorpha* Müll. Arg., which contains galbinic and norstictic acids. In the Southwest one may find *U. scabiosa* Mot (K−, P−, usnic acid only) which has a rugose cortex and numerous short fibrils or isidia.

111b Branches lacking angular ridges. 112

112a (111) Cortex eroding away on the main branches; isidia lacking. Fig. 425. ....
.... *Usnea longissima* **(L.) Ach.**

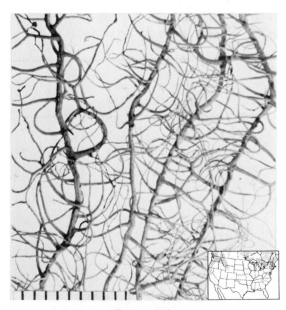

Figure 425

Figure 425   Usnea longissima

Thallus pale greenish yellow, pendulous, 15-30 cm long; main branches rugose, the cortex disappearing with age, the secondary branches with sparse soralia; apothecia rare. Medulla K−, C−, P− (usnic and diffractaic acids). Rare on branches of conifers. This classical lichen is much less common than *U. cavernosa* or *U. trichodea*. There are other chemical variants but little is known of them.

112b **Cortex continuous, strongly papillate with isidia developing.** ....
.... *Usnea ceratina* (see page 219)

113a (107) **Thallus pale green with fine, hair-like branches about 0.1 mm thick (use hand lens and ruler), without internal layers. Fig. 426.** ....
.... *Coenogonium interplexum* Nyl.

Figure 426

Figure 426   Coenogonium interplexum

Thallus dull yellowish green, forming mats 1-3 cm wide, the branches up to 0.5 cm long; algal component *Trentepohlia;* apothecia common, pale yellow, spores colorless, uniseriate,

1-septate, 2-3 x 6-9 μm. Common at the base of trees in mature forests. There are four other common species in this genus which must be separated with a microscope: *C. moniliforme* Tuck. has branches divided like a chain of pearls; *C. interpositum* Nyl. has simple biseriate spores; *C. linkii* Ehrenb. has a shelf-like thallus with 1-septate biseriate spores; and *C. implexum* Nyl. has uniseriate spores 2-4 x 7-12 μm.

113b **Thallus yellowish green, stiff, the branches 1 mm or more in diameter with a cortex and medulla.** 114

114a (113) **Central cord absent (section with razor blade toward base as shown in Fig. 427).** 115

114b **Dense central cord present.** 117

115a (114) **Surface strongly papillate; fibrils present. Fig. 427.** ....
.... *Usnea antillarum* (Vain.) Zahlbr.

Figure 427

Figure 427   Usnea antillarum

Thallus greenish yellow, tufted, rather stiff, basal branches inflated, 5-10 cm long; branches papillate, becoming densely sorediate-isidiate and isidiate; medulla turning reddish; apothecia lacking. Medulla K+ yellowish (unknown substances). Rare on trees in open woods. The hollow branches may not be noticed at first and the specimens will probably be identified as *U. comosa* or *U. mutabilis*. Two rarer southern Usneas have hollow branches: *U. implicita* Stirt. (base not inflated) and *U. vainioi* Mot. (sorediate without isidia; diffractaic acid present).

115b Surface without papillae, smooth or pseudocyphellate-striate. 116

116a (115) Isidiate soredia present; collected in the Pacific Northwest. Fig. 428. .... *Alectoria imshaugii* **Brodo & Hawks.**

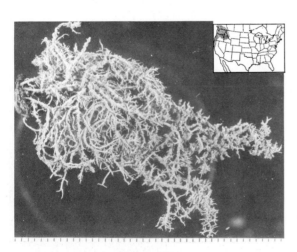

Figure 428

Figure 428   Alectoria imshaugii

Thallus yellowish green, tufted to somewhat pendulous, richly branched, 5-10 cm long;

branches rather stiff; apothecia lacking. Cortex K−; medulla K−, C−, P− or P+ yellow (usnic and squamatic or thamnolic acids). Widespread on conifers in open woods and along roads. This species can be easily mistaken for *Usnea subfloridana* at a distance but it is clearly hollow and has striate pseudocyphellae. Another confusable species, *Ramalina thrausta* (Ach.) Nyl., has soralia on apical hook-shaped branches; it is rarely collected in the boreal forests across southern Canada.

116b Isidia lacking; collected in the southeastern states. .... **(p. 211)** *Ramalina willeyi*

117a (114) Soredia and isidia lacking (use lens) (there may be numerous short fibrils); apothecia common. 118

117b Soredia and/or isidia present; apothecia rare. 119

118a (117) Collected in eastern United States. Fig. 429. .... *Usnea strigosa* **(Ach.) Eaton**

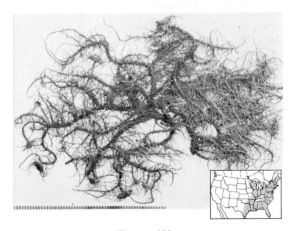

Figure 429

Figure 429   Usnea strigosa

Thallus greenish yellow, tufted, 3-8 cm long; branches moderately papillate; medulla rusty red or rose, especially toward the tips of branches, or pigment lacking; apothecia very common. Medulla K− or K+ red, P− or P+ yellow or orange (usnic acid alone or with norstictic and galbinic acids or psoromic or fumarprotocetraric acids). Common on canopy branches of deciduous trees and exposed trunks. There are several species in this variable group which must be identified with TLC. *Usnea subfusca* Stirt. has salazinic acid and strong papillae (as in Fig. 434) and occurs in the Appalachians. Both *U. evansii* Mot. and *U. tristis* are minor variants of *U. strigosa*. The psoromic acid population is found only in the southern states.

118b Collected in southwestern United States and California. Fig. 430. .... *Usnea arizonica* Mot.

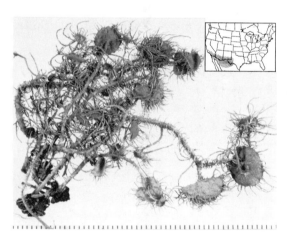

Figure 430

Figure 430   Usnea arizonica

Thallus pale greenish yellow, tufted, rather stiff, 3-6 cm long; main branches smooth to papillate, sometimes turning reddish but the medulla white; apothecia large, to 10 mm wide, the disk whitish pruinose. Cortex K−; medulla K+ yellow turning red, P+ orange (usnic and salazinic acids). This is one of the few easily identified Usneas in western United States.

119a (117) Medulla rusty red or rose (expose with razor blade). Fig. 431. .... *Usnea mutabilis* Stirt.

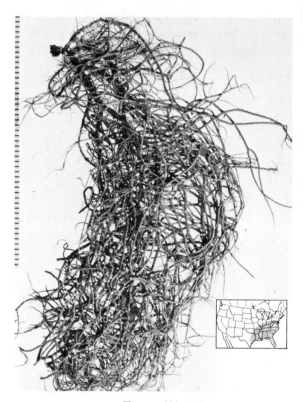

Figure 431

Figure 431   Usnea mutabilis

Thallus greenish yellow, tufted, 2-8 cm long; branches moderately papillate and isidiate-spinulate, the papillae becoming sorediate; apothecia lacking. Cortex and medulla K−, P− (usnic acid only). Common on deciduous trees in open woods. Because of the reddish medulla this species can be positively identified.

119b Medulla uniformly white (surface of cortex may be reddish). **120**

120a (119) Soredia powdery in orbicular soralia; isidia lacking. Fig. 432. .... ***Usnea fulvoreagens* (Räs.) Räs.**

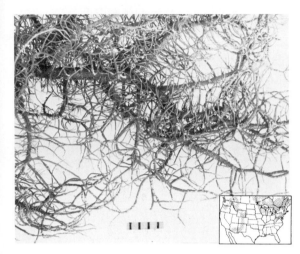

Figure 432

Figure 432   Usnea fulvoreagens

Thallus greenish yellow, tufted, 3-10 cm long; branches papillate, with large soralia; apothecia rare. Medulla K+ yellow, P+ orange (salazinic acid). Common on trees in open woods and along roadsides. The chemistry is variable, and some specimens may contain norstictic or stictic acids instead of salazinic. There are a number of microspecies (*U. betulina* Mot., *U. laricina* Vain.) but it is beyond the scope of this book to separate these.

120b Soredia becoming isidiate or only isidia present. **121**

121a (120) Branches without papillae, relatively long and sparsely branched and densely isidiate. Fig. 433. .... ***Usnea hirta* (L.) Wigg.**

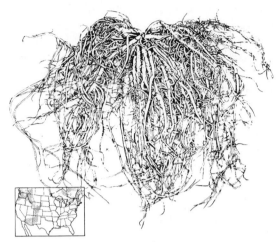

Figure 433

Figure 433   Usnea hirta (×1)

Thallus dark greenish yellow, tufted, flabby, 3-6 cm long; branches mostly smooth, papillae rare or lacking, densely fibrillose, in part sorediate-isidiate, the isidia long; apothecia lacking. Cortex and medulla K−, P− (usnic acid only). Common on hardwoods and conifers in open areas. The species is not well defined and will be confused with *U. subfloridana*. *U. variolosa* Mot. is reported to differ in having very short isidia but it is probably not distinct.

121b Branches with conspicuous papillae (see Fig. 434), soredia and isidia sparsely to moderately developed. **122**

Figure 434

Figure 434　Papillae on branches of *Usnea* sp. (×10)

122a (121) Branches turning dull reddish (especially at the base). Fig. 435. .... *Usnea rubicunda* Stirt.

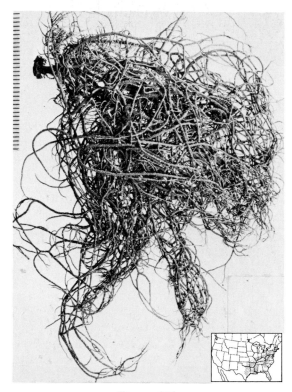

Figure 435

Figure 435　*Usnea rubicunda*

Thallus dark greenish yellow to rusty red, tufted, 2-6 cm long; branches moderately papillate, the papillae becoming sorediate and isidiate-sorediate; apothecia lacking. Medulla K+ yellow, P+ yellow (stictic acid). Common on tree trunks and rocks in mature forests. The reddish color and presence of stictic acid immediately identify this common *Usnea*.

122b Branches not turning reddish. 123

123a (122) Base of thallus more or less constricted, often blackening. Fig. 436. .... *Usnea subfloridana* Stirt.

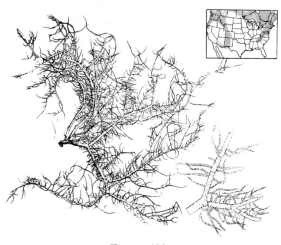

Figure 436

Figure 436　*Usnea subfloridana* (×1)

Thallus greenish yellow, tufted, 3-8 cm long; branches finely papillate, sorediate; apothecia lacking. Cortex K−; medulla K+ yellow turning red or K−, P+ orange or P− (usnic and norstictic, protocetraric, or thamnolic acids). Very common in trees in open woods. This is an extremely variable species with several chemical populations. Future research will undoubtedly show that a number of species are included in this complex.

123b Base of thallus not conspicuously constricted or blackened. 124

124a (123) Thallus growing on rocks. Fig. 437. .... *Usnea herrei* Hale

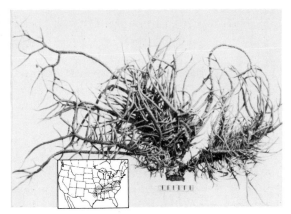

Figure 437

Figure 437   Usnea herrei

Thallus greenish yellow, rather stiff; branches short, with long fibrils, tending to fall in one direction, the cortex becoming reticulately rugose, papillate, the papillae sorediate and isidiate-sorediate; apothecia lacking. Medulla K+ yellow→red, P+ orange (norstictic acid). Common on exposed acidic rocks. The branching pattern and chemistry set this species apart. It is the commonest saxicolous *Usnea* in southeastern United States. However, you will collect other species on rock which do not fit here and which cannot be identified in our present state of knowledge.

124b Thallus growing on trees. 125

125a (124) Main branches coarsely papillate and sorediate; ultimate branches short and stubby. Fig. 438. ....
.... *Usnea ceratina* Ach.

Figure 438

Figure 438   Usnea ceratina (×3)

Thallus tufted to quite pendulous, basally attached or rarely free growing on branches; apothecia lacking. Cortex K−; medulla K−, C−, P− (diffractaic acid and usnic acid). Widespread on tree trunks in open woods. This species seems fairly distinct because of its chemistry and large size, 6-20 cm long. Large pendulous specimens from California have been called *U. californica* Herre and may be distinct from *U. ceratina*.

125b Main branches with smaller papillae and soralia; ultimate branches slender. Fig. 439. .... *Usnea dasypoga* (Ach.) Nyl.

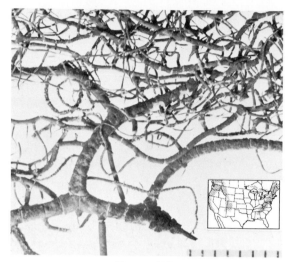

Figure 439

Figure 439   Usnea dasypoga

Thallus greenish yellow, tufted to almost pendulous, 6-30 cm long; sorediate sometimes isidiate; apothecia lacking. Cortex K−; medulla K+ yellow turning red, P+ orange (usnic and salazinic acids). Widespread on trees in open woods. This may well prove to represent a group of several closely related species when monographic studies are made.

126a (75) Branches flattened, sometimes with a distinct upper and lower surface. .. 127

126b Branches round in cross section. 136

127a (126) Collected in coastal California. 128

127b Collected in eastern North America and/or the western states. 131

128a (127) Thallus sorediate. 129

128b Thallus lacking soredia. 130

129a (128)Thallus soft and flabby with a whitish lower surface. .... (p. 206) *Evernia prunastri*

129b Thallus leathery, the upper and lower surfaces the same, white. Fig. 440. .... *Roccella babingtonii* Mont.

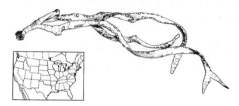

Figure 440

Figure 440   Roccella babingtonii (×1)

Thallus white to light mineral gray, 4-8 cm long, leathery, tufted to pendulous on trees or rocks; surface sorediate, the soralia capitate; apothecia very rare. Cortex and soredia C+ red (lecanoric acid). Locally abundant on exposed tree branches and rocks. Steeped in ammonia, this lichen is a good source of a deep purple dye.

130a (128) Surface of lobes C−, P+ red. Fig. 441. .... *Dendrographa leucophaea* (Tuck.) Darb.

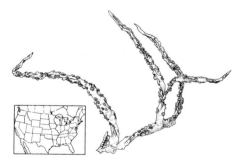

Figure 441

Figure 441   Dendrographa leucophaea (×1)

Thallus whitish mineral gray, 6-20 cm long, stiff and brittle, pendulous on branches or rock; apothecia common. Cortex and medulla P+ deep red (protocetraric acid). Common in exposed areas. *D. minor* Darb. has the same chemistry but is much smaller (3-6 cm) with thin rounded branches.

130b Surface of branches C+ red, P—. Fig. 442. .... *Roccella fimbriata* Darb.

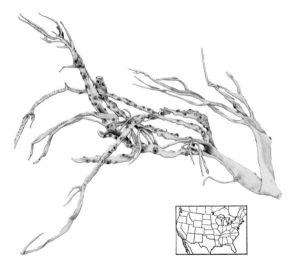

Figure 442

Figure 442   Roccella fimbriata (×1.5)

Thallus suberect to pendulous, leathery, grayish white, the lobes 1-4 mm wide; surface smooth, becoming rugose with age; apothecia common, the disk pruinose. Cortex K—, C+ red (lecanoric acid). Common on trees and shrubs in southern California and Baja California. Externally this is almost identical with *Dendrographa leucophaea* but the cortical structure and chemistry are different.

131a (127) Lower surface white and fibrous, lacking a cortex; cilia conspicuous (examine with hand lens). 132

131b Lower surface corticate, white, tan, or black; cilia lacking. 133

132a (131) Upper cortex K—; collected in northern United States and Canada. .... (p. 143) *Anaptychia setifera*

132b Upper cortex K+ yellow; collected in the southern parts of the United States. .. .... (p. 103) *Heterodermia leucomelaena*

133a (131) Collected in eastern North Amercia. .... (p. 96, nonisidiate) *Pseudevernia cladonia* or (p. 88, isidiate) *P. consocians.*

133b Collected in western North America. 134

134a (133) Lower surface channelled (see Fig. 481); medulla C+ red. .... *Pseudevernia intensa* (see page 97)

134b Lower surface more or less flat to convex; medulla C—. 135

135a (134) Margins of lobes sorediate; thallus soft, the surface dull, rugose. .... *Evernia prunastri* (see page 206)

135b Soredia lacking; thallus rather stiff, smooth and shiny. .... (p. 88, isidiate) *Platismatia herrei* and (p. 96, nonisidiate) *P. stenophylla.*

136a (126) Branches densely covered with numerous lobulelike phyllocladia (use hand lens and compare with Fig. 443); surface of branches often tomentose (Fig. 443). 137

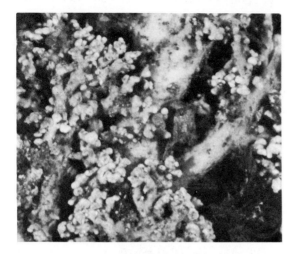

Figure 443

Figure 443 Phyllocladia and tomentum on pseudopodetia of *Stereocaulon tomentosum*

136b Branches (or stalks) smooth, lacking phyllocladia and tomentum. 145

137a (136) Pseudopodetia minute, 0.3-1.0 cm high, the phyllocladia granular or powdery; cephalodia lacking. Fig. 444. .... *Leprocaulon albicans* (Th. Fr.) Nyl.

Figure 444

Figure 444 Leprocaulon albicans

Pseudopodetia white, fragile; phyllocladia granular, decorticate, cephalodia lacking. Medulla K+ yellow, P+ yellow (atranorin, psoromic acid, or thamnolic acid). Rare on soil and in rock crevices. A similar species, *L. microscopicum* (Vill.) Gams, has a yellowish cast (usnic acid). *L. subalbicans* (Lamb) Lamb & Ward has no central branches but consists of a granular mass of phyllocladia. Both are western species.

137b Pseudopodetia coarse, more than 1 cm high, the phyllocladia distinct, corticated; cephalodia often present. 138

138a (137) Soredia present. Fig. 445. .... *Stereocaulon pileatum* Ach.

Figure 445

Figure 445 Stereocaulon pileatum (×2)

Pseudopodetia whitish mineral gray, erect, simple or sparingly branched; primary thallus persistent; soredia apical, capitate; apothecia common. Medulla KC+ reddish, P+ pale yellow (atranorin and lobaric acid). Common on stones and fence rows in open fields. In New England there is a related species, *S. nanodes* Tuck., which has an erect sorediate primary thallus. In the Olympic Mountains of Washington one may find *S. spathuliferum* Vain., which has a poorly developed or evanescent primary thallus.

138b Soredia lacking. **139**

139a (138) Growing directly on rock, usually firmly attached. **140**

139b Growing on soil or humus, more or less loosely attached. **141**

140a (139) Phyllocladia all cylindric-coralloid (use hand lens). Fig. 446. ....
.... *Stereocaulon dactylophyllum* **Flk.**

Figure 446

Figure 446 Stereocaulon   dactylophyllum (×1)

Pseudopodetia mineral gray, tufted, 2-6 cm tall; phyllocladia dense, tomentum thin or lacking; apothecia common, terminal. Medulla K+ yellow, P+ yellow orange (atranorin and stictic acid). Common on rocks in open outcrops. This species is frequently collected in the southern Appalachians. *S. intermedium* (Sav.) Magn., a rare species in the Pacific Northwest, is close except for chemistry (P−, lobaric acid).

140b Phyllocladia flattened, squamulose or digitate-squamulose. Fig. 447. ....
.... *Stereocaulon saxatile* **Magn.**

Figure 447

Figure 447 Stereocaulon saxatile

Pseudopodetia whitish mineral gray, loosely attached to adnate, 3-6 cm tall, often forming orbicular colonies; phyllocladia short, dense, the tomentum sparse to thick, gray; cephalodia sparse; apothecia common, terminal. Medulla KC+ reddish, P+ pale yellow (atranorin and lobaric acid). Common on boulders and stone fences in open areas. This species is frequently collected in the northern states. It intergrades broadly with *S. paschale,* which has numerous cephalodia. From New England to Nova Scotia it may occur with *S. glaucescens* Tuck., which has granulose to crenate-squamulose phyllocladia.

141a (139) Primary thallus persistent as small squamules or granular crusts; pseudopodetia short, erect, 1-2 cm high. 142

141b Primary thallus evanescent or lacking; pseudopodetia taller, 2-6 cm long, prostrate or erect. 143

142a (141) Cephalodia blackish brown, scabrid (use hand lens). Fig. 448. .... *Stereocaulon condensatum* **Hoffm.**

Figure 448

Figure 448   Stereocaulon condensatum (×1.5)

Pseudopodetia mineral gray, covering extensive areas of sandy soil, 1-2 cm high; phyllocladia dense, digitate, tomentum; sparse; apothecia common, terminal. Medulla KC+ reddish, P+ pale yellow (atranorin and lobaric acid). Common on soil in fields and open areas. This is the only common *Stereocaulon* so closely attached to soil in eastern North America.

142b Cephalodia reddish brown, smooth. Fig. 449. .... *Stereocaulon glareosum* **(Sav.) Magn.**

Figure 449

Figure 449   Stereocaulon glareosum (×3)

Pseudopodetia whitish mineral gray, 1-2 cm high, traces of primary thallus sometimes present; phyllocladia flattened, short, tomentum often well developed, whitish rosy; apothecia common, terminal. Medulla KC+ reddish, P+ pale yellow (atranorin and lobaric acid). Widespread on sandy soil in open areas. *S. incrustatum* Flk., a rare species in Colorado, differs in having thick gray tomentum in which the scattered phyllocladia are partially immersed.

143a (141) Tomentum on the branches conspicuous and continuous. Fig. 450. .... *Stereocaulon tomentosum* **Fr.**

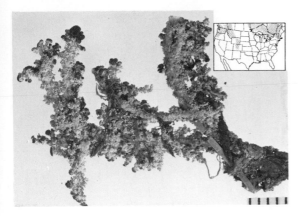

Figure 450

Figure 450   Stereocaulon tomentosum

Pseudopodetia whitish mineral gray, growing loosely on the substratum, prostrate to sub-erect, 4-8 cm tall; phyllocladia dense, digitate; cephalodia small, concealed in the tomentum; apothecia common, lateral. Medulla K+ yellow, P+ pale orange (atranorin and stictic acid). Common on soil over rocks and on humus. The thick tomentum and numerous lateral apothecia characterize this species. A chemical variant with lobaric acid (KC+ reddish) occurs in western North America with the typical form. Another western species with stictic acid, S. *myriocarpum* Th. Fr., differs in having large black cephalodia (1 mm across) and more appressed tomentum. It occurs from Washington and Montana southward into Mexico.

143b **Tomentum sparse or lacking, surface of branches bare.** 144

144a (143) **Cephalodia numerous, small, blackish, scabrid (use hand lens); pseudopodetia erect. Fig. 451.** .... *Stereocaulon paschale* (L.) Hoffm.

Figure 451

Figure 451   Stereocaulon paschale (×2)

Pseudopodetia whitish mineral gray, tufted or growing loosely on the substratum, 4-8 cm tall; phyllocladia dense, short and squamiform; apothecia common, mostly terminal. Medulla KC+ red, P+ pale yellow (atranorin and lobaric acid). Widespread on soil and among mosses. The range of this species overlaps that of S. *saxatile,* which has sparse inconspicuous cephalodia.

144b **Cephalodia rare, brownish, smooth; pseudopodetia decumbent. Fig. 452.** .... *Stereocaulon rivulorum* Magn.

Figure 452

Figure 452   Stereocaulon rivulorum (×2)

Pseudopodetia whitish mineral gray, forming low dense colonies, 2-4 cm broad; phyllocladia small, almost granular, cephalodia sparse; apothecia rare. Medulla KC+ red, P+ pale yellow (atranorin and lobaric acid). Rare on rock and soil in arctic regions.

145a (136) Surface of branches dull, fibrous, lacking a cortex (use hand lens and see Fig. 392); thallus richly branched. .. 146

145b Surface of branches corticate, usually shiny (ecorticate but not fibrous in *Baeomyces*); thallus usually unbranched (richly branched only in *Cladonia furcata* and *Sphaerophorus globosus*). .. 147

146a (145) Thallus forming compact heads without a distinct main stem. Fig. 453. .... *Cladina evansii* (Abb.) Hale & Culb.

Figure 453

Figure 453   Cladina evansii (×½)

Thallus (podetia) growing loosely on white sand, 3-5 cm high, often covering extensive areas, white; branches finely divided; pycnidia and apothecia not uncommon. Medulla K−, C−, P− (perlatolic acid). Very common on sandy soil in scrubby areas. This is one of the most conspicuous lichens along the Atlantic and Gulf coasts and in Florida.

146b Thallus forming scattered entangled masses without discrete heads but with a distinct main stem. Fig. 454. .... *Cladina rangiferina* (L.) Harm.

Figure 454

Figure 454 Cladina rangiferina (×1)

Podetia 6-10 cm high, the colonies often extensive; branching pattern mostly in fours, the axils open; pycnidia common, apothecia very rare, dark brown. K+ yellow (atranorin), P+ red (fumarprotocetraric acid). Common on soil and humus in open areas. This well-known lichen will usually occur with yellowish species as *C. subtenuis* and *C. arbuscula* and by holding up the species together in bright light or sunshine, the ashy gray color can be distinguished from the pale yellowish gray of the other two.

147a (145) Thallus richly branched. **148**

147b Thallus unbranched or sparsely branched. **150**

148a (147) Squamules present on branches (examine carefully with hand lens). ......
.... **(p. 191)** *Cladonia furcata*

148b Squamules lacking. **149**

149a (148) Collected on trees or rocks in boreal and western North America. Fig. 455. ....
.. *Sphaerophorus globosus* **(Huds.) Vain.**

Figure 455

Figure 455 Sphaerophorus globosus

Thallus greenish gray to tan, erect to subpendulous, 4-8 cm long, rather stiff; surface of branches smooth and shiny; apothecia spherical, at tips of branches, the disc sooty black. Cortex and medulla K−, C−, P+ yellow (sphaerophorin, squamatic acid, thamnolic acid, and hypothamnolic acid). Common on fir and spruce in the humid Cascade forests and on soil and humus in the arctic regions. The overall aspect reminds one of *Cladonia* but the apothecia are unique. A second species, *S. fragilis* (L.) Pers., will be found only in arctic areas as far south as New England and can be separated by a test with iodine (in solution with potassium iodide): *S. fragilis* is I−, while *S. globosus* is I+ blue.

149b Collected on soil in the Coastal Plain. ....
.... **(p. 203)** *Cladonia pachycladodes*

150a (147) Thallus hollow (use hand lens and section with razor blade). **151**

150b Thallus (stalks) solid. **153**

151a (150) Thallus greenish mineral gray, arising from a crustose primary thallus and firmly attached to soil. Fig. 456. .... Pycnothelia papillaria (Ehrh.) Duf.

Figure 456

Figure 456   Pycnothelia papillaria (×2)

Podetia light mineral gray, 0.5-1.5 cm tall, irregularly inflated and becoming branched, often constricted at the base; primary thallus granular, scattered; pycnidia common, apothecia rare. Medulla K—, C—, P— (fatty acids). Widespread on soil along roads and other open areas. This species was formerly classified in *Cladonia*.

151b Thallus yellowish tan or white, growing loosely on soil in arctic-alpine habitats. 152

152a (151) Thallus white, pointed, wormlike. Fig. 457. .... Thamnolia subuliformis (Ehrh.) Culb.

Figure 457

Figure 457   Thamnolia subuliformis (×1)

Thallus 3-6 cm long, with some side branches, erect or prostrate; pycnidia and apothecia unknown. K+ yellowish, P+ yellow (baeomycic acid). Very common on exposed soil and among mosses. There is hardly a place in the Arctic where one will not find the scattered thalli of this unique lichen. A chemical variant, *T. vermicularis* (Sw.) Schaer., makes up one-third to one-half of the specimens collected; it is K+ deep yellow (thamnolic acid).

152b Thallus pale yellowish tan, blunt-tipped, fingerlike. .... (p. 202) Dactylina arctica

153a (150) Collected on shrubs and rocks in southern California. Fig. 458. .... Schizopelte californica Th. Fr.

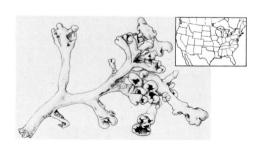

Figure 458

Figure 458   Schizopelte californica (×2)

Thallus grayish white, rather brittle, 3-6 cm across, the branches coarsely inflated, 1-2 mm thick, sparsely branched; surface dull; apothecia common. Cortex and medulla K−, C+ red, P− (lecanoric acid). Common on rocks and shrubs in the fog zone. This is another of the curious California-Baja California endemics that include *Dendrographa*, *Roccella*, *Niebla*, and other genera.

153b Collected on rocks and soil outside of southern California. 154

154a (153) Apothecia black. Fig. 459. ....
.... *Pilophoron aciculare* (Ach.) Nyl.

Figure 459

Figure 459   Pilophoron aciculare (×1.5)

Stalks (pseudopodetia) light mineral gray, slender, 1-3 cm tall; primary thallus granular to minutely areolate; apothecia common, round. Cortex K+ yellow, C−, P+ faint yellow (atranorin and zeorin). Common on rocks in the moist Cascade forests. Another species in the West, *P. hallii* (Tuck.) Vain., has elongate rather than round apothecia. In alpine districts of eastern North America one may find *P. cereolus* (Ach.) Nyl., which has small stalks less than 1 cm tall and round apothecia.

154b Apothecia pink or flesh-colored to tan. ..
... 155

155a (154) Apothecia pink, large and orbicular. Fig. 460. ....
.... *Baeomyces fungoides* (Sw.) Ach.

Figure 460

Figure 460   Baeomyces fungoides (×1.5)

Pseudopodetia white, 0.3-0.6 cm tall, unbranched; primary thallus crustose, granular, whitish mineral gray; apothecia common, rounded, light pink. P+ yellow (baeomycic acid). Common on soil along open roadbanks. The primary crust consolidates freshly exposed soil and soon produces the attractive pink apothecia. The species was formerly called *B. roseus*.

155b Apothecia small and flattened, flesh-colored to tan or pinkish. 156

156a (155) Thallus very thin, greenish. Fig. 461. .... *Baeomyces absolutus* Tuck.

Figure 461

Figure 461   Baeomyces absolutus

Pseudopodetia white, 0.1-0.3 cm high; primary thallus thin, greenish, smooth; apothecia pink. P+ yellow (baeomycic acid). Rare on rocks in mature forests. *B. rufus* (Huds.) Rebent., common on rocks and soil in northern and western

United States, is similar but has a grayish thallus and reacts K+ yellow (stictic acid).

156b Thallus thicker, granular to subsquamulose. Fig. 462. ....
.... *Baeomyces carneus* (Retz.) Flk.

Figure 462

Figure 462   Baeomyces carneus

Pseudopodetia brownish mineral gray, 0.2-0.5 cm tall; primary thallus forming a rather thick crust, 3-8 cm broad; apothecia common, peltate, dark brown. K+, P+ yellow (norstictic acid). Rare on boulders in open forests. This species is inconspicuous and difficult to collect from flat granitic rocks. In arctic regions one will find *B. placophyllus* Ach., which has a very thick almost lobate thallus.

V. SQUAMULOSE LICHENS

This is an artificial grouping of lichens which share a squamulose growth form. Two genera make up the bulk of the species treated here: the primary thalli of *Cladonia* which lack podetia and *Psora*, a genus formerly included in *Lecidea*. This key makes no pretense of enabling one to identify all sterile *Cladonia* squamules that will be collected. Only the common species that normally lack podetia are discussed below. In addition, the taxonomy of *Psora* (and *Dermatocarpon*) has not been critically revised and the treatment here is also provisional. Wetmore's monograph can be consulted for a full account of the Heppiaceae (*Heppia* and *Peltula*), especially by students living in the arid western states where they are so common.

1a   Squamules yellowish, whitish, or greenish gray. 2

1b   Squamules pink, reddish, tan, brown, or blackish. 10

2a   (1) Squamules adnate on soil, contiguous. Fig. 463. ....
.... *Squamarina lentigera* (Web.) Poelt

Figure 463

Figure 463 Squamarina lentigera

Squamules crowded, 1-3 mm wide, with a vague white rim, fragile, forming colonies 3-5 cm broad; apothecia common, the disc tan to light orangish brown. Cortex K+ yellow (atranorin). Widespread on calcareous soils, especially gypsum, in exposed areas.

2b   Squamules suberect or ascending, loosely attached to free growing. 3

3a   (2) Squamules with a yellowish green cast above and yellow or cream-colored below. .... (p. 196) *Cladonia robbinsii* and (p. 197) *C. strepsilis*

3b   Squamules whitish to pale greenish gray, white or gray below. 4

4a   (3) Squamules K+ yellow or yellow turning red (test lower surface). 5

4b   Squamules K–. 8

5a   (4) Squamules growing loosely on white sand, separate, curling into balls. Fig. 464. .... *Cladonia prostrata* Evans

Figure 464

Figure 464 Cladonia prostrata (×1)

Squamules very large, up to 3 cm long, curled upward at the margins when dry, the lower surface chalky white; podetia lacking but pyc-

nidia common. K+ yellow (atranorin) and P+ red (fumarprotocetraric acid). Widespread on exposed beaches. This could be confused with no other *Cladonia*.

5b   Squamules more or less attached to soil or rock, crowded, not curling. 6

6a   (5) Squamules long and narrow (1-2 cm) with a blackened base; collected only in the Great Smoky Mountains. Fig. 465. .... *Gymnoderma lineare* (Evans) Yosh. & Sharp

Figure 465

Figure 465 Gymnoderma lineare (×1)

Squamules dark greenish mineral gray; lower surface white to brownish toward the tips, weakly corticated; podetia lacking but small clustered apothecia common on lobe tips. K+ yellow (atranorin). This is one of the most unusual endemic lichens in North America and should not be collected by individuals. It was formerly called a *Cladonia*.

6b   Squamules about 1 cm long or less, the base not blackened; collected throughout North America. 7

7a   (6) Squamules K+ yellow turning red (norstictic acid). .... (p. 193) *Cladonia polycarpoides*

7b Squamules K+ persistent yellow (norstictic acid lacking). ....
.... (p. 193) *Cladonia symphycarpa* and (p. 192) *C. turgida*

8a (4) Apothecia present on surface of squamules. ....
.... (p. 234) *Psora decipiens*

8b Apothecia lacking. 9

9a (8) Squamules chalky white below. ......
.... (p. 196) *Cladonia apodocarpa* and *C. subapodocarpa*

9b Squamules dull white to grayish below.
.... (p. 183) *Cladonia mateocyatha*

10a (1) Collected on tree bark or rocks. .. 11

10b Collected on soil, soil in rock crevices, or on burned tree stumps. 15

11a (10) Squamules 5 mm or more wide; black dots (perithecia) present (use hand lens). Fig. 466. .... *Dermatocarpon tuckermanii* (Rav.) Zahlbr.

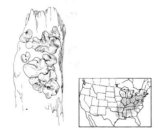

Figure 466

Figure 466 Dermatocarpon tuckermanii (×1)

Thallus light brown, adnate, 2-6 cm broad and composed of separate lobes (actually squamules) 3-6 mm long; lower surface brown, bare; upper surface with black dots (perithecia). Widespread but not commonly collected on deciduous trees, especially white oak, in open woods. This is the only species of *Dermatocarpon* in North America that grows on trees.

11b Squamules 0.5-1 mm wide; apothecia present. 12

12a (11) Apothecia with a conspicuous lobulate margin. Fig. 467. ....
.... *Psoroma hypnorum* (Vahl.) S. Gray

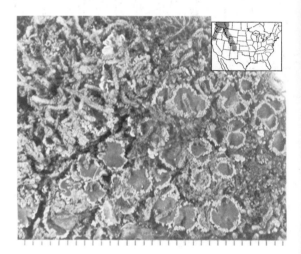

Figure 467

Figure 467 Psoroma hypnorum

Thallus light brown, consisting of fine squamules 0.5-1 mm long forming colonies 3-6 cm broad, brown below and corticate; apothecia numerous, 2-4 mm in diameter with a thick, finely lobulate rim; disc brown; spores colorless, simple. Common on soil and over mosses along trails in open woods.

12b   Apothecia with a smooth margin. 13

13a   (12) Squamules lacking isidia. Fig. 468. .... *Pannaria leucosticta* (Tuck.) Nyl.

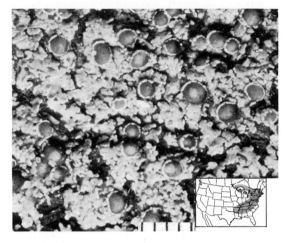

Figure 468

Figure 468   Pannaria leucosticta

Thallus composed of numerous confluent squamules, 3-6 cm broad; margins white pruinose with raised finger-like projections; lower surface rather dark, tomentose; apothecia common, the rim white. On deciduous trees, rarely on rocks, in mature forests. It is inconspicuous and not commonly collected. This species is typical of a series of nonsorediate, nonisidiate Pannarias, including *P. leucostictoides* Ohlss., a pruinose species in the Pacific Northwest, *P. crossophylla* (Tuck.) Merr. & Burnh., which lacks a distinct thalline rim on the apothecia and has finely dissected, nearly cylindrical squamules, and *P. leucophaea* (Vahl) Jørg., which also has a poorly developed thalline rim and flat imbricate squamules. *Pannaria rubiginosa*, which was keyed out under the foliose

lichens (see page 143), may also key out here if the lobes are interpreted as squamules.

13b   Squamules isidiate. 14

14a   (13) Collected on limestone and other calcareous rocks. .... (p. 150) *Placynthium nigrum*

14b   Collected on trees. Fig. 469. .... *Parmeliella tryptophylla* (Ach.) Müll. Arg.

Figure 469

Figure 469   Parmeliella tryptophylla (×12)

Thallus brown, consisting of squamules about 1 mm wide with digitate isidia on the margins; lower surface pale with bluish black tomentum; apothecia rare, red brown, without a thalline margin. Widespread on trees in open woods.

15a   (10) Squamules sorediate (use hand lens). Fig. 470. .... *Psora scalaris* (Ach.) Hook.

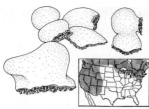

Figure 470

Figure 470   Psora scalaris (×10)

Squamules about 1 mm long, forming crowded colonies several cm wide, somewhat ascending and hoop-shaped (under lens), sorediate on the lower surface; apothecia rare. Medulla C+, KC+ red (lecanoric acid). Common on dead wood and charred stumps. Another similar species but without the C+ reaction is *P. anthracophila* (Nyl.) Arn. which has the same habitat. Another member of this group, *P. friesii* (Ach.) Hellb., lacks soredia and also reacts C−.

15b   Squamules lacking soredia. **16**

16a   (15)   Upper surface with black dots (perithecia); apothecia never present. Fig. 471. .... ***Dermatocarpon lachneum*** **(Ach.) A. L. Sm.**

Figure 471

Figure 471   Dermatocarpon lachneum

Squamules dark brown, 1-3 mm wide or more, rather closely adnate on soil, crowded, forming colonies 3-8 cm broad; perithecia common. Common on soil among boulders or in rock crevices. Calcareous soils are definitely preferred by this extremely variable lichen. In desert regions it consolidates the soil. Several species in this group have been reported from North America but all probably belong to *D. lachneum.*

16b   Upper surface with round, adnate apothecia. **17**

17a   (16)   Squamules pink or brick red, often becoming white pruinose; apothecia marginal. Fig. 472. ....
.... ***Psora decipiens*** **(Ach.) Hoffm.**

Figure 472

Figure 472   Psora decipiens

Squamules adnate on soil, 2-4 mm wide, forming colonies 2-8 cm broad; rim of squamules white pruinose but not raised; apothecia black, convex. Widespread on soil, especially in arid regions. This is an extremely common and variable species, ranging from pure white to reddish or pink.

17b Squamules brown to blackish, lacking pruina. **18**

18a (17) Squamules ascending with a whitish lower surface, resembling a *Cladonia*. Fig. 473. .... ***Psora novomexicana* Lesd.**

Figure 473

Figure 473   Psora novomexicana

Squamules brown, large and ascending, thin, crowded, to 4 mm wide, the margin white-rimmed; lower surface ivory-colored, appearing pruinose; apothecia black. On soil over rocks, often in crevices. Widespread in the western states. This is one of the easier Psoras to recognize in the field because of the large *Cladonia*-like squamules.

18b Squamules more appressed to adnate without a contrasting white lower surface. **19**

19a (18) Apothecia to 1.5 mm broad, immersed in the upper surface. Fig. 474. .... ***Heppia lutosa* (Ach.) Nyl.**

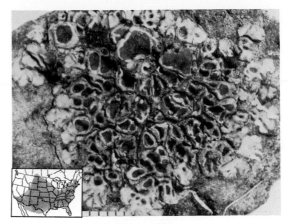

Figure 474

Figure 474   Heppia lutosa

Squamules brown to greenish olive, 2-6 mm wide, scattered to more or less organized into a lobate colony; algae blue-green; apothecia common. Widespread on calcareous soil but often overlooked. This strange genus is most common in semidesert regions of western North America. Closely related *Peltula* has 16-100 spores per ascus rather than 8 as in *Heppia,* but these two genera are externally very similar and occur together. Wetmore's monograph should be consulted for more details.

19b Apothecia 1 mm or less in diameter, adnate on the squamules. Fig. 475. .... ***Psora russellii* (Tuck.) Schneid.**

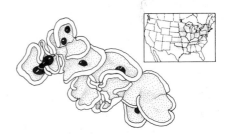

Figure 475

Figure 475 *Psora russellii* (×10)

Squamules brown, adnate to slightly ascending at the margins, 2-3 mm wide, forming colonies 1-4 cm broad; rim of squamules white; apothecia black. Widespread on soil in open areas and deserts. This species is representative of a whole series of very poorly known Psoras. For example, *P. icterica* (Mont.) Müll. Arg., has a yellow medulla, *P. rubiformis* (Wahl.) Hook. has thick, white-rimmed squamules with strongly convex apothecia, and *P. rufonigra* (Nyl.) Schneid. has a uniformly dark brown thallus with crowded, overlapping squamules and plane apothecia. Keys in the regional studies by Weber and Wetmore should be consulted for further help in this difficult group.

List of Synonyms and Incorrect Names

The names of many lichens continue to change because taxonomic research has shown them to be incorrect and a new unfamiliar name must be adopted. This is an unfortunate but necessary step in botanical progress. In addition, a number of large genera have recently been split up into smaller, more homogeneous genera, and these names will also cause some initial confusion. The following list gives the more important names used in the first edition of *How to Know the Lichens* and the equivalent names used in this edition. Recent checklists by Hale & Culberson (see the chapter on Useful References) will give complete lists of name changes.

Alectoria abbreviata = Bryoria abbreviata
Alectoria americana = Bryoria trichodes
Alectoria fremontii = Bryoria fremontii
Alectoria fuscescens = Bryoria fuscescens
Alectoria nidulifera = Bryoria furcellata
Alectoria pseudofuscescens = Bryoria pseudofuscescens
Alectoria pubescens = Pseudephebe pubescens
Anaptychia casarettiana = Heterodermia casarettiana
Anaptychia diademata = Heterodermia diademata
Anaptychia echinata = Heterodermia echinata
Anaptychia granulifera = Heterodermia granulifera
Anaptychia hypoleuca = Heterodermia hypoleuca
Anaptychia kaspica = Anaptychia setifera
Anaptychia leucomelaena = Heterodermia leucomelaena
Anaptychia obscurata = Heterodermia obscurata
Anaptychia ravenelii = Heterodermia albicans
Anaptychia rugulosa = Heterodermia rugulosa
Anaptychia speciosa = Heterodermia speciosa
Anaptychia squamulosa = Heterodermia squamulosa

Caloplaca murorum = Caloplaca saxicola
Cladonia alpestris = Cladina stellaris
Cladonia arbuscula = Cladina arbuscula
Cladonia evansii = Cladina evansii
Cladonia mitis = Cladina mitis
Cladonia nemoxyna = Cladonia rei
Cladonia papillaria = Pycnothelia papillaria
Cladonia rangiferina = Cladina rangiferina
Cladonia subtenuis = Cladina subtenuis
Coccocarpia parmelioides = Coccocarpia erythroxyli
Collema subfurvum = Collema subflaccidum
Cornicularia aculeata = Coelocaulon aculeatum
Lecanora chrysoleuca = Rhizoplaca chrysoleuca
Leptogium hirsutum = Leptogium burnetiae
Pannaria pityrea = Pannaria conoplea
Parmelia amazonica = Pseudoparmelia amazonica
Parmelia arnoldii = Parmotrema arnoldii
Parmelia atticoides = Neofuscelia atticoides
Parmelia aurulenta = Parmelina aurulenta
Parmelia austrosinensis = Parmotrema austrosinense
Parmelia baltimorensis = Pseudoparmelia baltimorensis

Parmelia caperata = Pseudoparmelia caperata
Parmelia caroliniana = Pseudoparmelia caroliniana
Parmelia centrifuga = Xanthoparmelia centrifuga
Parmelia cetrata = Parmotrema cetratum
Parmelia chlorochroa = Xanthoparmelia chlorochroa
Parmelia congruens = Pseudoparmelia sphaerospora
Parmelia confoederata = Bulbothrix confoederata
Parmelia conspersa = Xanthoparmelia conspersa
Parmelia crinita = Parmotrema crinitum
Parmelia cristifera = Parmotrema cristiferum
Parmelia crozalsiana = Pseudoparmelia crozalsiana
Parmelia cryptochlorophaea = Pseudoparmelia cryptochlorophaea
Parmelia cumberlandia = Xanthoparmelia cumberlandia
Parmelia dilatata = Parmotrema dilatatum
Parmelia dissecta = Parmelina dissecta
Parmelia endosulphurea = Parmotrema endosulphureum
Parmelia eurysaca = Parmotrema eurysacum
Parmelia formosana = Hypotrachyna formosana
Parmelia galbina = Parmelina galbina
Parmelia horrescens = Parmelina horrescens
Parmelia hypopsila = Xanthoparmelia hypopsila
Parmelia hypotropa = Parmotrema hypotropum
Parmelia imbricatula = Hypotrachyna imbricatula
Parmelia laevigatula = Bulbothrix laevigatula
Parmelia livida = Hypotrachyna livida
Parmelia loxodes = Neofuscelia loxodes
Parmelia mellissii = Parmotrema mellissii
Parmelia michauxiana = Parmotrema michauxianum
Parmelia novomexicana = Xanthoparmelia novomexicana
Parmelia obsessa = Parmelina obsessa
Parmelia perforata = Parmotrema perforatum
Parmelia plittii = Xanthoparmelia plittii
Parmelia praesorediosa = Parmotrema praesorediosum
Parmelia prolongata = Hypotrachyna prolongata
Parmelia quercina = Parmelina quercina
Parmelia rampoddensis = Parmotrema rampoddense
Parmelia reticulata = Parmotrema reticulatum

Parmelia revoluta = Hypotrachyna revoluta
Parmelia rockii = Hypotrachyna rockii
Parmelia rutidota = Pseudoparmelia rutidota
Parmelia salacinifera = Pseudoparmelia salacinifera
Parmelia scortella = Bulbothrix goebelii
Parmelia sinuosa = Hypotrachyna sinuosa
Parmelia stuppea = Parmotrema stuppeum
Parmelia subisidiosa = Parmotrema subisidiosum
Parmelia subramigera = Xanthoparmelia subramigera
Parmelia subtinctoria = Parmotrema subtinctorium
Parmelia sulphurata = Parmotrema sulphuratum
Parmelia taractica = Xanthoparmelia taractica
Parmelia tasmanica = Xanthoparmelia tasmanica
Parmelia texana = Pseudoparmelia texana
Parmelia tinctorum = Parmotrema tinctorum
Parmelia ultralucens = Parmotrema ultralucens
Parmelia xanthina = Parmotrema xanthinum
Parmeliopsis halei = Parmeliopsis subambigua
Physcia adiastola = Phaeophyscia adiastola
Physcia albicans = Physcia crispa
Physcia ciliata = Phaeophyscia ciliata
Physcia constipata = Phaeophyscia constipata
Physcia endococcina = Phaeophyscia decolor
Physcia grisea = Physconia detersa
Physcia lacinulata = Phaeophyscia imbricata
Physcia muscigena = Physconia muscigena
Physcia pulverulenta = Physconia pulverulenta
Physcia pusilloides = Phaeophyscia pusilloides
Physcia rubropulchra = Phaeophyscia rubropulchra
Physcia sciastra = Phaeophyscia sciastra
Physcia setosa = Phaeophyscia hispidula
Physcia syncolla = Physciopsis syncolla
Physcia tribacoides = Physcia americana
Ramalina combeoides = Niebla combeoides
Ramalina fastigiata = Ramalina americana
Ramalina homalea = Niebla homalea
Ramalina minuscula = Fistulariella dilacerata
Ramalina roesleri = Fistulariella roesleri
Rinodina oreina = Dimelaena oreina
Stereocaulon albicans = Leprocaulon albicans
Umbilicaria papulosa = Lasallia papulosa
Usnea comosa = Usnea subfloridana

Phylogenetic List of Genera and Families

The genera treated in this book are arranged below by order and family. Under each family the genera are listed alphabetically. It should be remembered, however, that lichenologists are not in full agreement on the classification of lichens and that the family units given below are subject to modification. See a recent Hale and Culberson Checklist for further details.

CLASS ASCOMYCETES

Order Caliciales
Sphaerophoraceae: *Sphaerophorus*.

Order Lecanorales
Suborder Lecanorineae
Acarosporaceae: *Acarospora*.

Alectoriaceae: *Alectoria, Bryoria, Coelocaulon, Cornicularia,* and *Pseudephebe*.

Anziaceae: *Anzia*.

Baeomycetaceae: *Baeomyces*.

Candelariaceae: *Candelaria* and *Candelina*.

Cladoniaceae: *Cladina, Cladonia, Pilophoron,* and *Pycnothelia*.

Coccocarpiaceae: *Coccocarpia*.

Collemataceae: *Collema* and *Leptogium*.

Heppiaceae: *Heppia* and *Peltula*.

Hypogymniaceae: *Cavernularia, Hypogymnia,* and *Menegazzia*.

Lecanoraceae: *Agrestia, Lecanora,*
Omphalodium, Placopsis, and *Rhizoplaca*.

Lecideaceae: *Psora*.

Pannariaceae: *Pannaria, Parmeliella,* and *Psoroma*.

Parmeliaceae: *Bulbothrix, Cetraria, Cetrelia, Everniastrum, Hypotrachyna, Neofuscelia, Parmelia, Parmelina, Parmeliopsis, Parmotrema, Platismatia, Pseudevernia, Pseudoparmelia,* and *Xanthoparmelia*.

Ramalinaceae: *Fistulariella, Niebla,* and *Ramalina*.

Stereocaulaceae: *Leprocaulon* and *Stereocaulon*.

Umbilicariaceae: *Lasallia* and *Umbilicaria*.

Usneaceae: *Dactylina, Evernia, Letharia, Thamnolia,* and *Usnea*.

Suborder Lichinineae
Lichinaceae: *Ephebe*.

Suborder Peltigerineae
Placynthiaceae: *Placynthium*.
Peltigeraceae: *Hydrothyria, Nephroma, Peltigera,* and *Solorina*.

Stictaceae: *Lobaria, Pseudocyphellaria,* and *Sticta*.

Suborder Teloschistineae
Teloschistaceae: *Caloplaca, Teloschistes,* and *Xanthoria*.

Suborder Physciineae
Physciaceae: *Anaptychia, Dimelaena, Dirinaria, Heterodermia, Phaeophyscia, Physcia, Physciopsis, Physconia,* and *Pyxine*.

Order Gyalectales
Gyalectaceae: *Coenogonium*.

Order Verrucariales
Verrucariaceae: *Dermatocarpon*.

Order Arthoniales
Roccellaceae: *Dendrographa, Roccella,* and *Schizopelte*.

Acknowledgments
for Illustrations

Abbayes, H. des: 483, 487, 488 (Traité de Lichenologie, LeChevalier, 1951).

Asahina, Y.: 315, 338 (Illustrated Flora of the Japanese Cryptogams, 1939).

Frey, E.: 48, 445, 448, 449, 452 (Rabenhorst Kryptogamenflora, vol. 9, 1933, 1934).

Grassi, M.: 16, 478 (Lilloa, vol. 25, 1950).

Hale, M. E.: 1, 19 (Biology of Lichens, London, 1967).

Halliday, N.: 81, 83, 125, 168, 173, 313, 337, 340, 344, 347, 352, 354, 356, 358, 359, 361, 363-65, 369-71, 375-78, 380, 388, 397, 401-03, 412-414, 457, 459-62, 464-66, 481, 482, 489, 495, 497, 502, 505, 506 (originals).

Harris, C. W.: 420, 424 (Bryologist, vol. 4, 1900).

Henssen, A.: 325 (Symbolae Bot. Ups. 18, 1963).

Howard, G.: 58, 60, 391, 399, 400 (Bryologist, vol. 66, 1963).

Lindahl, P. O.: 507 (Svensk Bot. Tidskr, vol. 47, 1953).

Menez, E.: 4, 35, 44, 102, 148, 157, 163, 166, 179, 183, 196, 207, 280, 407, 470, 475, 477, 480, 490, 491 (originals).

Motyka, J.: 46, 259, 260, 265, 276, 324, 327, 433, 436 (Flora Polska: Porosty (Lichenes), vol. 5, 1960, 1962).

Riddle, L. W.: 496 (Bot. Gaz. vol. 50, 1910).

Schneider, A.: 291, 476, 493, 498, 500, 501 (A Textbook of Lichenology, 1897).

Schroeder, J.: 18, 27, 37, 42, 55-57, 59, 61-66, 73, 77, 79, 88, 94, 99, 100, 101, 103, 109, 118, 122, 133, 144, 147, 159, 161, 172, 177, 178, 193, 204, 219, 221, 251-53, 256, 270, 271, 282, 288, 298, 300, 302, 314, 317, 320, 323, 336, 367, 406, 410, 418 (originals).

Schumacher, E.: 54, 113 (originals).

Tangerini, A.: 104, 105, 128, 175, 189, 192, 273, 343, 353, 372, 442, 458, 484 (originals).

Tateoka, S.: 47, 111, 127, 139 (originals).

Trass. H.: 339, 341, 342, 345, 346, 348, 350, 351, 357, 373, 374, 383, 384, 395, 396, 405, 454 (Lood. Selts. Eesti Tead. Juures, vol. 5, 1958).

Index and
Pictured Glossary

A

ABORTIVE: imperfect or poorly developed, as podetia in some Cladonias.

Acarospora
  chlorophana, 32

ACICULAR: long and needle-shaped, as spores. Fig. 476.

Figure 476 (X400)

ADNATE: lying flat on and attached to the substratum. See Fig. 3.

Agrestia
  hispida, 201

Alectoria
  fallacina, 212
  imshaugii, 215
  lata, 212
  nigricans, 212
  ochroleuca, 212
  sarmentosa, 212
  vancouverensis, 212

ALGAL LAYER: a thin layer of green or blue-green algae lying just below the upper cortex of stratified lichens. See Fig. 8.

Anaptychia
  palmatula, 140
  setifera, 143

ANNULATE: ringed, referring to the cortex of Usneas. Fig. 477.

Figure 477 (X2)

Anzia
  colpodes, 94
  ornata, 79

B

APICAL: at the terminal part of a lobe or podetium.

APOTHECIUM: the reproductive structure of ascomycetes containing the hymenium (asci, spores, and paraphyses, usually disc- or cup-shaped). See Fig. 15.

APPRESSED: lying flat on and firmly attached to the substratum. See Fig. 3.

AQUATIC: growing in water or partially submerged.

AREOLATE: composed of areoles, as the surface of some Cladonias or the thallus of some crustose lichens.

AREOLES: individual segmented parts of an areolate thallus.

ARTICULATED: jointed or segmented, as the branches of some Usneas. See Fig. 477.

ASCUS: a sac (20-100 μm long) containing spores. Fig. 478.

Figure 478 (X200)

AXIL: the upper angle between branches of fruticose lichens, either open or closed.

Baeomyces
  absolutus, 229
  carneus, 230
  fungoides, 229
  placophyllus, 230
  rufus, 229

C

Bryoria
  abbreviata, 173
  capillaris, 173
  fremontii, 172
  friabilis, 172
  furcellata, 171
  fuscescens, 172
  glabra, 172
  lanestris, 172
  nadvornikiana, 173
  oregana, 173
  pseudofuscescens, 174
  tortuosa, 174
  trichodes, 174

BULBATE: inflated, as the basally inflated cilia of *Bulbothrix.* Fig. 479.

Figure 479 (X5)

Bulbothrix
  confoederata, 100
  coronata, 100
  goebelii, 91
  laevigatula, 91

CALCAREOUS: referring to rocks containing calcium or lime.

Caloplaca
  saxicola, 29
  scopularis, 30
  trachyphylla, 30

CANALICULATE: grooved on the lower surface. See Fig. 481.

Candelaria

concolor, 35
  fibrosa, 46

Candelina
  mexicana, 32
  submexicana, 32

CAPITATE: shaped like a head, referring to soralia. Fig. 480.

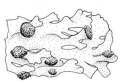

Figure 480 (X2)

Cavernularia
  hultenii, 84
  lophyrea, 98

CEPHALODIA: tiny thalli (0.5-1.0 mm) growing on the upper cortex (or even internally) in *Peltigera, Placopsis, Lobaria,* and *Stereocaulon.* See Fig. 6G.

Cetraria
  arenaria, 169
  aurescens, 46
  chlorophylla, 120
  ciliaris, 133
  coralligera, 128
  cucullata, 198
  culbersonii, 126
  fendleri, 139
  halei, 134
  hepatizon, 145
  idahoensis, 95
  islandica, 169
  merrillii, 138
  nivalis, 198
  oakesiana, 36
  orbata, 134
  pallidula, 45
  pinastri, 35
  platyphylla, 133
  sepincola, 139
  tilesii, 198
  viridis, 46
  weberi, 139

Cetrelia

cetrarioides, 57
chicitae, 57
monachorum, 57
olivetorum, 56
CHANNELED: grooved, as the lower surface of *Pseudevernia*. Fig. 481.

Figure 481 (X1)

CHINKY: cracked and fissured, as in some curstose lichens. See Fig. 39.
CILIA: hairlike outgrowths along margins of lobes. Fig. 482.

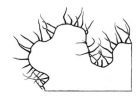

Figure 482 (X3)

CILIATE: provided with cilia.
Cladina
  arbuscula, 200
  evansii, 226
  impexa, 201
  mitis, 200
  pacifica, 201
  rangiferina, 226
  stellaris, 199
  submitis, 201
  subtenuis, 199
  tenuis, 200
  terrae-novae, 201
Cladonia
  abbreviatula, 184
  acuminata, 189
  amaurocraea, 202
  apodocarpa, 196
  atlantica, 184
  bacillaris, 186
  bacilliformis, 188
  balfourii, 188
  beaumontii, 195
  bellidiflora, 184
  boryi, 203
  botrytes, 194
  brevis, 194
  caespiticia, 195
  calycantha, 180
  capitata, 194
  caraccensis, 182
  cariosa, 193
  carneola, 177
  caroliniana, 203
  cenotea, 179

chlorophaea, 178
clavulifera, 193
coccifera, 176
coniocraea, 187
conista, 178
cornuta, 190
crispata, 183
cristatella, 184
cryptochlorophaea, 178
cyanipes, 189
cylindrica, 188
decorticata, 189
deformis, 177
didyma, 186
digitata, 176
dimorphoclada, 204
ecmocyna, 181
farinacea, 191
fimbriata, 178
floerkeana, 186
floridana, 184
furcata, 191
glauca, 191
gonecha, 177
gracilis, 181
grayi, 178
hypoxantha, 185
incrassata, 185
leporina, 201
macilenta, 185
macrophylla, 194
major, 178
mateocyatha, 183
merochlorophaea, 178
multiformis, 182
norrlinii, 189
ochrochlora, 188
pachycladodes, 203
parasitica, 187
perlomera, 178
phyllophora, 181
piedmontensis, 192
pityrea, 189
pleurota, 177
polycarpia, 193
polycarpoides, 193
prostrata, 231
pseudomacilenta, 187
pseudorangiformis, 191
pyxidata, 180
rappii, 180
ravenelii, 185
rei, 179
robbinsii, 196
santensis, 195
scabriuscula, 190
squamosa, 182
strepsilis, 197
subapodocarpa, 197
subrangiformis, 191
subsetacea, 204
subsquamosa, 183
subulata, 191
symphycarpa, 193
transcendens, 184
turgida, 192
uncialis, 204
verticillata, 180
COALESCE: fuse together, as many thalli merging into a single large colony.
Coccocarpia
  cronia, 90
  erythroxyli, 95
Coelocaulon
  aculeatum, 170
Coenogonium
  implexum, 214

interplexum, 214
interpositum, 214
linkii, 214
moniliforme, 214
Collema
  bachmanianum, 153
  callibotrys, 158
  coccophorum, 153
  conglomeratum, 158
  crispum, 150
  cristatum, 150
  flaccidum, 150
  fragrans, 158
  furfuraceum, 157
  leptaleum, 158
  limosum, 153
  multipartitum, 152
  nigrescens, 156
  occultatum, 158
  polycarpon, 152
  pulcellum, 157
  ryssoleum, 151
  subflaccidum, 157
  tenax, 152
  tuniforme, 151
  undulatum, 151
COLONY: a group of lichen thalli growing closely together.
CONTIGUOUS: touching or in close contact, as lobes.
CORALLOID: resembling coral, as richly branched isidia. Fig. 483.

Figure 483 (X5)

CORD: a dense strand of hyphae in the center of branches of *Usnea*.
CORIACEOUS: leathery and not easily broken and crumbled.
Cornicularia
  californica, 139
  divergens, 170
  normoerica, 171
CORTEX: the outermost layer of the thallus, consisting of compressed hyphal cells appearing cellular (paraplectenchyma) or as parallel strands (prosoplectenchyma). See Fig. 8.
CORTICATE: having a distinct cortex.
CORTICOLOUS: growing on tree trunks or branches.
CRENATE: having a notched edge, as some lobe margins.
CRUSTOSE: a lichen growth form with thalli growing in intimate contact with the substratum and lacking a lower cortex and rhizines.
CYPHELLAE: large circular, recessed pores in the lower surface of *Sticta*. See Fig. 86.

D

Dactylina
  artica, 202
  madreporiformis, 202
  ramulosa, 202
DECORTICATE: lacking a cortex, leaving the medulla directly exposed.
Dendrographa
  leucophaea, 220
  minor, 221
DENTATE: a finely toothed edge, as in some lobe margins.
Dermatocarpon
  fluviatile, 132
  lachneum, 234
  miniatum, 163
  moulinsii, 163
  reticulatum, 163
  tuckermanii, 232
DICHOTOMOUS: dividing into two parts, as the forking branching pattern in foliose and fruticose thalli and in rhizines. Fig. 484.

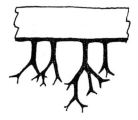

Figure 484 (X10)

DIFFUSE: scattered or without definite pattern, as diffuse soredia.
DIGITATE: arranged like fingers, as in phyllocladia of *Stereocaulon*.
Dimelaena
  oreina, 32
Dirinaria
  aspera, 92
  confusa, 99
  frostii, 85
  picta, 84
  purpurascens, 100
DISC: the surface of an apothecium.
DISCRETE: separate and distinct, as lobes or thalli.
DISSECTED: cut up or divided into many lobes or lobules.
DORSIVENTRAL: having distinct upper and lower surfaces that are different in appearance.

E

ECILIATE: lacking cilia.
ECORTICATE: lacking a cortex, as the lower surface of *Cladonia* squamules and thalli of some *Heterodermias*. See Fig. 191.
ELLIPSOIDAL: shaped like an

ellipse, usually referring to spores.

ENTIRE: smooth and unbroken, referring to lobe margins.

Ephebe
  americana, 170
  lanata, 170
  solida, 170

ESOREDIATE: lacking soredia.
EVANESCENT: disappearing at maturity, as the primary thallus of *Cladina*.

Evernia
  divaricata, 205
  mesomorpha, 205
  prunastri, 206

Everniastrum
  catawbiense, 83

EXPANDED: spreading out, as the thalli of large foliose lichens.

F

FARINOSE: powdery as flour, referring to soredia.
FIBRILLOSE: having fibrils.
FIBRILS: short lateral branches, almost isidialike, as in *Usnea*. Fig. 485.

Figure 485 (X2)

FIBROUS: composed of fibers and appearing cottony, as the lower surface of some species of *Heterodermia*. See Fig. 7.
FILAMENTOUS: a lichen growth form, thin, hairlike strands of filamentous algae and fungi as in *Coenogonium* and *Ephebe*.
FISSURED: deeply cracked, as in the cortex of some foliose and fruticose lichens. See Fig. 6A.
Fistulariella
  dilacerata, 210
  inflata, 211
  roesleri, 205
FOLIOSE: a lichen growth form, leaflike.
FOVEOLATE: finely pitted or dimpled, referring to the thallus surface.
FRUTICOSE: a lichen growth form, shrubby or hairlike. See Fig. 10.
FURCATE: regularly forked, referring to branching patterns of lobes and podetia.
FUSIFORM: spindle-shaped, used to describe spores.

G

GELATINOUS: like jelly or gelatin, used to describe unstratified lichens containing blue-green algae when wet.
GRANULAR: grainlike, used to describe coarse soredia.
Gymnoderma
  lineare, 231

H

HAIRS: fine multicellular outgrowths from the cortex, as in *Phaeophyscia cernohorskyi* (Fig. 231).
HAPTER: a tuft or loosely aggregated hyphae on the lower surface in *Collema*.
Heppia
  lutosa, 235
Heterodermia
  albicans, 105
  appalachensis, 103
  casarettiana, 104
  diademata, 115
  echinata, 110
  erinacea, 111
  granulifera, 110
  hypoleuca, 111
  leucomelaena, 102
  obscurata, 104
  pseudospeciosa, 106
  rugulosa, 114
  speciosa, 105
  squamulosa, 111
Hydrothyria
  venosa, 147
HYPHAE: microscopic multicellular fungal threads that make up the lichen thallus. Fig. 486.

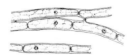

Figure 486 (X400)

Hypogymnia
  appalachensis, 73
  austerodes, 74
  bitteri, 73
  duplicata, 75
  enteromorpha, 76
  heterophylla, 75
  imshaugii, 74
  inactiva, 75
  krogii, 74
  metaphysodes, 75
  occidentalis, 76
  oroarctica, 99
  pseudophysodes, 74
  rugosa, 89
  tubulosa, 73
  vittata, 73
Hypotrachyna
  croceopustulata, 82
  dentella, 92
  ensifolia, 92
  formosana, 82
  gondylophora, 83
  imbricatula, 92
  laevigata, 83
  livida, 101
  oostingii, 83
  producta, 83
  prolongata, 92
  pulvinata, 98
  pustulifera, 82
  revoluta, 83
  rockii, 83
  showmanii, 82
  sinuosa, 36
  thysanota, 83
  virginica, 101

I

IMPERFORATE: lacking holes, used in describing closed axils of branches or discs of apothecia.
INCISED: deeply notched, as margins of lobes or squamules.
IRREGULAR: uneven, used to describe irregularly broadened lobes.
ISIDIA: fingerlike cylindrical outgrowths from the upper cortex. Fig. 487.

Figure 487 (X10)

ISIDIOID: isidialike growths on lichen thalli.

L

LABRIFORM: lip-shaped, usually referring to apical soralia. Fig. 488.

Figure 488 (X3)

LACERATE: with jagged edges or tips, as in lobe margins and podetia.
LACINIATE: divided into numerous small lobes.
LAMELLAE: thin plates, referring to acid crystals.
LAMINAL: superficial on the surface of the thallus, as soredia or apothecia.
Lasallia
  papulosa, 163
  pensylvanica, 163
Lecanora
  muralis, 33
  novomexicana, 33
Lepraria
  finkii, 7
Leprocaulon
  albicans, 222
  microscopicum, 222
  subalbicans, 222
Leptogium
  apalachense, 152
  arsenei, 155
  austroamericanum, 154
  azureum, 155
  burgessii, 148
  burnetiae, 148
  californicum, 150
  chloromelum, 153
  corticola, 156
  crenatellum, 156
  cyanescens, 154
  dactylinum, 154
  denticulatum, 154
  digitatum, 149
  floridanum, 155
  isidisellum, 153
  juniperinum, 156
  laceroides, 148
  lichenoides, 149
  marginellum, 153
  microstictum, 156
  millegranum, 155
  palmatum, 151
  papillosum, 148
  phyllocarpum, 155
  platynum, 151
  rivale, 148
  rugosum, 149
  saturninum, 148
  sessile, 155
  sinuatum, 151
  stipitatum, 155
  tenuissimum, 150
Letharia
  columbiana, 198
  vulpina, 197
LINEAR: narrow with a uniform width, as lobes or soralia. Fig. 489.

Figure 489 (X1)

Lobaria
  hallii, 118
  linita, 130
  oregana, 44
  pulmonaria, 117
  quercizans, 132
  ravenelii, 131
  scrobiculata, 118
  tenuis, 132
LOBATE: forming lobes, referring usually to the thallus margin of some crustose lichens. See Fig. 39.
LOBE: a rounded or strap-shaped division of the thallus. Fig. 490.

Figure 490 (X2)

LOBULATE: provided with numerous marginal or laminal lobules.
LOBULE: subdivision of a lobe.

M

MARGINAL: located at or along lobe margins, as soralia. Fig. 491.

Figure 491 (X2)

MARKINGS: white reticulate or spotted outlines on the surface of lobes. See Fig. 6C.
Massalongia
   carnosa, 125
MEDULLA: inner part of the thallus consisting of loosely interwoven hyphae. See Fig. 8.
MEMBRANE: a thin covering over the cup in podetia of *Cladonia*.
Menegazzia
   terebrata, 72
MICROCONIDIA: uninucleate bacilliform cells produced in pycnidia. Fig. 492.

Figure 492

MOTTLED: variegated white and black or brown, as on the lower surface of some foliose lichens.
MURIFORM: spores divided into many chambers by transverse and longitudinal walls. Fig. 493.

Figure 493 (X200)

N

Neofuscelia
   ahtii, 146
   atticoides, 145
   brunella, 146
   chiricahuensis, 129
   infrapallida, 146
   loxodes, 128
   occidentalis, 146
   subhosseana, 128
   verruculifera, 128
Nephroma
   arcticum, 40
   bellum, 134
   helveticum, 134
   laevigatum, 135
   parile, 120
   resupinatum, 55, 131
Niebla
   cephalota, 209
   ceruchis, 209
   combeoides, 208
   homalea, 208
NODULOSE: having nodules, small knots in the branches.
NONISIDIATE: not having any isidia.
NONSOREDIATE: not having any soredia.
NOSTOC: a blue-green alga which forms filaments. Fig. 494.

Figure 494 (X400)

O

Omphalodium
   arizonicum, 159

P

Pannaria
   ahlneri, 123
   conoplea, 123
   crossophylla, 233
   leucophaea, 233
   leucosticta, 233
   leucostictoides, 233
   lurida, 141
   rubiginosa, 142
   tavaresii, 127

PAPILLAE: small rounded bumps on the cortex. Figs. 434 and 495.

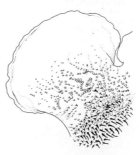

Figure 495 (X5)

PAPILLATE: having papillae.
PARAPHYSES: threadlike hyphae packing the spaces between asci in the hymenium. See Fig. 16.
PARAPLECTENCHYMATOUS: cortical tissue apearing as packed cells observed with a microscope.
Parmelia
   albertana, 121
   appalachensis, 58
   bolliana, 58
   borreri, 56
   darrovii, 45
   disjuncta, 126
   elegantula, 129
   exasperata, 136
   exasperatula, 129
   flaventior, 34
   fraudans, 78
   glabra, 135
   glabratula, 119
   glabroides, 135
   halei, 139
   infumata, 129
   multispora, 136
   olivacea, 136
   omphalodes, 97
   panniformis, 129
   perreticulata, 56
   praesignis, 45
   pseudosulcata, 89
   reddenda, 57
   rudecta, 58
   saxatilis, 89
   septentrionalis, 137
   sorediosa, 125
   sphaerosporella, 45
   squarrosa, 89
   stictica, 120
   stygia, 145
   subargentifera, 121
   subaurifera, 119
   subelegantula, 129
   subolivacea, 135
   subpraesignis, 38
   subrudecta, 56
   substygia, 125
   sulcata, 78
   trabeculata, 137
   ulophyllodes, 35
Parmeliella
   pannosa, 142
   plumbea, 142
   tryptophylla, 233

Parmelina
   aurulenta, 80
   dissecta, 93
   galbina, 100
   horrescens, 93
   obsessa, 87
   quercina, 98
Parmeliopsis
   aleurites, 110
   ambigua, 36
   capitata, 37
   hyperopta, 107
   placorodia, 116
   subambigua, 37
Parmotrema
   arnoldii, 62
   austrosinense, 65
   cetratum, 70
   crinitum, 68
   cristiferum, 64
   diffractaicum, 60
   dilatatum, 63
   dominicanum, 64
   endosulphureum, 65
   eurysacum, 71
   hababianum, 62
   haitiense, 67
   hypoleucinum, 60
   hypoleucites, 59
   hypotropum, 60
   internexum, 68, 94
   louisianae, 60
   madagascariaceum, 38
   margaritatum, 62
   mellissii, 67
   michauxianum, 70
   neotropicum, 65
   perforatum, 70
   perlatum, 63
   praeperforatum, 70
   praesorediosum, 64
   rampoddense, 61
   reticulatum, 60
   rigidum, 70
   robustum, 65
   simulans, 60
   stuppeum, 62
   subisidiosum, 67
   subtinctorium, 66
   sulphuratum, 37
   tinctorum, 66
   ultralucens, 68
   xanthinum, 38
PELTATE: umbrella-shaped.
Peltigera
   aphthosa, 48
   canina, 50
   collina, 49
   degenii, 51
   elizabethae, 52
   evansiana, 50
   horizontalis, 52
   lepidophora, 50
   leucophlebia, 49
   malacea, 52
   membranacea, 51
   polydactyla, 52
   praetextata, 51
   rufescens, 51
   scabrosa, 53
   spuria, 49
   venosa, 51
Peltula
   euploca, 160
PENDULOUS: hanging down or draping tree branches.
PERFORATE: pierced with holes.

PERITHECIA: flask-shaped fruiting bodies of pyrenocarpous lichens. See Fig. 309.

Phaeophyscia
  adiastola, 123
  cernohorskyi, 122
  ciliata, 140
  constipata, 146
  decolor, 146
  endococcinodes, 146
  erythrocardia, 140
  hirtella, 140
  hispidula, 124
  imbricata, 141
  orbicularis, 123
  pusilloides, 124
  rubropulchra, 118
  sciastra, 140

PHYLLOCLADIA: tiny granular or lobed leaflike structures on branches of Stereocaulon. Figs. 443 and 496.

Figure 496 (X5)

Physcia
  adscendens, 103
  aipolia, 114
  alba, 114
  albinea, 113
  americana, 106
  biziana, 115
  caesia, 107
  callosa, 108
  chloantha, 121
  clementi, 107
  crispa, 79
  dubia, 107
  halei, 112
  melanchra, 122
  millegrana, 101
  phaea, 113
  pseudospeciosa, 106
  sorediosa, 84
  stellaris, 116
  subtilis, 102
  tenella, 103

Physciopsis
  adglutinata, 122
  syncolla, 114

Physconia
  detersa, 119
  enteroxantha, 119
  muscigena, 144
  pulverulenta, 137

Pilophoron
  aciculare, 229
  cereolus, 229
  hallii, 229

Placopsis
  gelida, 85

Placynthium
  nigrum, 150

PLANE: flat and smooth, referring to surface of lobes.
PLATES: flattened rhizine-like structures on lower surface in Umbilicaria. See Fig. 309.

Platismatia
  glauca, 63

  herrei, 88
  lacunosa, 69
  norvegica, 66
  stenophylla, 96
  tuckermanii, 69

PODETIUM: a hollow simple or branched upright structure in Cladonia. Fig. 497.

Figure 497 (X1)

PRIMARY THALLUS: the squamulose or granular thallus of Cladonia.
PROLIFERATE: to produce parts in succession, as cups in some Cladonias.
PROSOPLECTENCHYMATOUS: referring to cortical tissue where cells periclinally arranged. Also seen in cord of Usnea.
PROSTRATE: lying flat on the substratum.
PRUINA: a fine white woolly or granular covering on the upper cortex or disc of apothecia. See Fig. 6D.

Pseudephebe
  minuscula, 171
  pubescens, 170

Pseudevernia
  cladonia, 96
  consocians, 88
  intensa, 97

PSEUDOCYPHELLATE: provided with pseudocyphellae, white pores in the upper or lower cortex. See Fig. 6B.

Pseudocyphellaria
  anomala, 55
  anthraspis, 55
  aurata, 55
  rainierensis, 56

Pseudoparmelia
  alabamensis, 82, 85
  amazonica, 94
  baltimorensis, 38
  caperata, 34
  caroliniana, 90
  crozalsiana, 80
  cryptochlorophaea, 81
  martinicana, 94
  rutidota, 47
  salacinifera ,109
  sphaerospora, 48
  texana, 82

PSEUDOPODETIUM: the upright fruticose thallus of Stereocaulon and Baeomyces. Fig. 498.

Figure 498 (X1)

Psora
  anthracophila, 234
  decipiens, 234
  friesii, 234
  icterica, 236
  novomexicana, 235
  rubiformis, 236
  rufonigra, 236
  russellii, 235
  scalaris, 233

Psoroma
  hypnorum, 232

PULVINATE: growing like small cushions.
PUNCTIFORM: shaped like small dots, as in tiny orbicular soralia.
PUSTULATE: covered with pustules, erupting blister-like structures. See Fig. 136.
PYCNIDIA: small flask-shaped reproductive structures in the medulla producing microconidia. See Fig. 15B.

Pycnothelia
  papillaria, 228

Pyxine
  caesiopruinosa, 86
  eschweileri, 78
  sorediata, 86

R

Ramalina
  americana, 210
  complanata, 210
  denticulata, 210
  ecklonii, 209
  evernioides, 206
  farinacea, 207
  hypoprotocetrarica, 207
  intermedia, 206
  leptocarpha, 209
  menziesii, 207
  montagnei, 211
  paludosa, 210
  petrina, 206
  pollinaria, 205
  sinensis, 210
  stenospora, 211
  subleptocarpha, 207
  tenuis, 211
  thrausta, 215
  usnea, 208
  willeyi, 211

RETICULATE: arranged as a network, as cracks in the upper cortex. See Fig. 6A.
REVOLUTE: rolled downward, as tips of sorediate lobes.
RHIZINATE: provided with rhizines.
RHIZINES: strands of hyphae on the lower surface of many foliose lichens. Fig. 499.

Figure 499 (X10)

Rhizoplaca
  chrysoleuca, 159
  melanophthalma, 159

RIMOSE: finely chinked or fissured, usually referring to crustose lichens.

Roccella
  babingtonii, 220
  fimbriata, 221

ROTUND: rounded in outline, as tips of broad lobes.
RUGOSE: having wrinkles or ridges, usually in upper cortex.

S

SAXICOLOUS: growing on rocks.
SCABRID: having fine white scales on the upper cortex. See Fig. 6D.

Schizopelte
  californica, 228

SEPARATE: not joined or in close contact, referring to patterns of lobe branching. See Fig. 59.
SEPTATE: divided into two or more parts by a septum or wall, as septate spores. Fig. 500.

Figure 500 (X400)

SESSILE: attached directly to the thallus without a stalk.

Solorina
  bispora, 144
  crocea, 51, 130
  octospora, 144
  saccata, 143

SORALIA: clumps of soredia on the surface or margins of thalli.
SOREDIA: microscopic clumps of several algal cells surrounded by hyphae and erupting at the thallus surface as a powder.

Sphaerophorus
  fragilis, 227
  globosus, 227

SPINULATE: provided with spinules.
SPINULES: short stiff branchlets resembling large isidia, most frequently seen in Usnea.

SPORES: microscopic reproductive cells of fungi contained in the ascus. Fig. 501.

Figure 501 (X400)

Squamarina
  lentigera, 230
SQUAMULATE: provided with squamules, as podetia of *Cladonia*. Fig. 502.

Figure 502 (X1)

SQUAMULE: a small scalelike thallus which lacks a lower cortex and rhizines. See Fig. 31.
SQUAMULOSE: lichen growth form, referring to thalli consisting of squamules as in *Cladonia*.
SQUARROSE: branching pattern of rhizines. Fig. 503.

Figure 503 (X10)

Stereocaulon
  condensatum, 224
  dactylophyllum, 223
  glareosum, 224
  glaucescens, 224
  incrustatum, 224
  intermedium, 223
  myriocarpum, 225
  nanodes, 223
  paschale, 225
  pileatum, 222
  rivulorum, 225
  saxatile, 223
  spathuliferum, 223
  tomentosum, 224
Sticta
  fuliginosa, 53
  limbata, 54
  sylvatica, 54
  weigelii, 54

STIGONEMA: a filamentous blue-green alga. Fig. 504.

Figure 504 (X400)

STRATIFIED: consisting of horizontal layers, referring to the internal structure of lichens which have a cortex, algal layer, and medulla. Fig. 505.

Figure 505 (X200)

STRIAE: elongate white ridges (0.1-1 mm long) in the cortex of some fruticose lichens.
SUBASCENDING: with tips of lobes rising somewhat above the substratum.
SUBCRUSTOSE: a growth form intermediate between crustose and foliose, the center of the thallus usually being crustose, the margins lobed.
SUBERECT: ascending strongly toward the thallus margin, referring to lichens intermediate between foliose and fruticose. See Fig. 3C.
SUBFRUTICOSE: a growth form intermediate between foliose and fruticose. See Fig. 390.
SUBISIDIATE: sparsely or imperfectly isidiate, the isidia sometimes intermingled with soredia.
SUBSQUAMULOSE: imperfectly squamulose.
SUBSTRATUM: the medium (soil, rock, bark, etc.) on which a lichen grows.

T

Teloschistes
  chrysophthalmus, 175
  exilis, 175
  flavicans, 175
THALLUS: plant body of a lichen, usually classified by growth form (crustose, foliose, and fruticose).
Thamnolia
  subuliformis, 228
  vermicularis, 228
TOMENTUM: a growth of woolly or felty hairs on the lower surface of the thallus and on pseudopodetia of *Stereocaulon*. See Fig. 7C.

TUBERCULAR: warty or knoblike, often applied to cortex of *Usnea*.

U

Umbilicaria
  angulata, 161
  caroliniana, 161
  cylindrica, 162
  decussatus, 165
  deusta, 160
  hirsuta, 160
  hyperborea, 166
  kraschennikovii, 164
  mammulata, 161
  muhlenbergii, 165
  phaea, 166
  polyphylla, 166
  polyrrhiza, 161
  proboscidea, 164
  torrefacta, 165
  vellea, 161
  virginis, 162
UMBILICATE: a growth form in which the thallus is attached to the substratum at the center by an umbilicus. See Fig. 29.
UMBILICUS: a single strand of rhizines on the lower surface of umbilicate lichens.
UNSTRATIFIED: lacking distinct layers, referring to the internal structure of gelatinous lichens, which lack separate algal and medullary layers. Fig. 506.

Figure 506 (X200)

Usnea
  angulata, 213
  antillarum, 214
  arizonica, 216
  betulina, 217
  californica, 219
  cavernosa, 213
  ceratina, 219
  dasypoga, 219
  dimorpha, 213
  evansii, 216
  fulvoreagens, 217
  herrei, 219
  hirta, 217
  implicita, 215
  laricina, 217
  longissima, 213
  mutabilis, 216
  rubicunda, 218
  scabiosa, 213
  strigosa, 215
  subfloridana, 218
  subfusca, 216
  trichodea, 212
  vainioi, 215
  variolosa, 217

V

VEINS: raised riblike structures on the lower surface

of *Peltigera* and *Solorina*. Fig. 507.

Figure 507 (X2)

VERRUCOSE: covered with wartlike growths.

W

WHITE-RETICULATE: having a netted pattern of white lines, as the lobe tips of some foliose lichens. See Fig. 6C.
WHITE-SPOTTED: having numerous tiny white spots on the upper cortex, as in *Physcia*. See Fig. 6E.

X

Xanthoparmelia
  ajoensis, 39
  arseneana, 42
  centrifuga, 41
  chlorochroa, 40
  congensis, 40
  conspersa, 39
  cumberlandia, 42
  dierytha, 40
  dissensa, 43
  huachucensis, 43
  hypomelaena, 43
  hypopsila, 43
  ioanis-simae, 42
  joranadia, 39
  kurokawae, 39
  lineola, 42
  mexicana, 40
  monticola, 42
  mougeotii, 36
  novomexicana, 41
  piedmontensis, 39
  plittii, 40
  psoromifera, 42
  subcentrifuga, 37
  subdecipiens, 42
  subramigera, 39
  taractica, 42
  tasmanica, 44
  tinctorum, 39
  tucsonensis, 42
  weberi, 39
  wyomingica, 41
Xanthoria
  candelaria, 30
  elegans, 29
  fallax, 30
  parietina, 31
  polycarpa, 31
  ramulosa, 31
  sorediata, 29